More praise for

Swimming with Piranhas at Feeding Time

"Delightfully addictive; read this one for natural-history fun."

—*Booklist*

"A delightful collection. . . . With warmth and simplicity, the author spins a beguiling web as he recalls his travels. . . . Bright entertainment from a great explainer of the lives of animals."

—*Kirkus Reviews*

"[Conniff] captures outlandish and obscure creatures with his eminently digestible prose—leavened with civilized wit and a well-developed sense of irony. . . . Whether he is tracking fearsome predators, or merely promiscuous dung beetles, Conniff transports the wild things right into our cozy dens. He makes us root harder for their survival in a world that is stacked against so many of them."   —David Holahan, *Christian Science Monitor*

"As entertaining as the title suggests . . . this is sure to inspire wanderlust."   —Jonathan Christison, *New Scientist*

"From big cats to horseshoe crabs and snapping turtles to termites, *Swimming with Piranhas at Feeding Time* is an adventurous and uncommon tour of the animal kingdom. . . . Richard Conniff writes with vibrancy and verve."

—Jason Weeks, *Sacramento Book Review*

Richard CONNIFF

Swimming with PIRANHAS
at Feeding Time

My Life Doing DUMB STUFF with Animals

W. W. NORTON & COMPANY

New York | *London*

For information about permission to reproduce selections from this
book, write to Permissions, W. W. Norton & Company, Inc.,
500 Fifth Avenue, New York, NY 10110

For information about special discounts for bulk purchases,
please contact W. W. Norton Special Sales at specialsales@wwnorton.com
or 800-233-4830

Manufacturing by Courier Westford
Book design by Judith Stagnitto Abbate/Abbate Design
Production manager: Devon Zahn

Library of Congress Cataloging-in-Publication Data

Conniff, Richard, 1951–
    Swimming with piranhas at feeding time : my life doing dumb stuff
with animals / Richard Conniff.
        p. cm.
    Includes bibliographical references.
    ISBN 978-0-393-06893-1 (hardcover)
    1. Dangerous animals—Anecdotes. 2. Animals—Anecdotes. 3. Conniff,
Richard, 1951—–Travel. I. Title.
QL100.C66 2009
590—dc22

                                                        2008051234

ISBN 978-0-393-30457-2 pbk.

W. W. Norton & Company, Inc.
500 Fifth Avenue, New York, N.Y. 10110
www.wwnorton.com

W. W. Norton & Company Ltd.
Castle House, 75/76 Wells Street, London W1T 3QT

1 2 3 4 5 6 7 8 9 0

For Karen, who had the sense to stay home

Contents

Contents

Swimming with PIRANHAS
at Feeding Time

Preface

When I'm heading off to report a story, people often say, "You're going *where*? You're going to do *what*?" After a little while, doubtfully, they add, "And somebody's actually paying you for this?" Then they ask if they can come along. No editor or producer has ever posted a job description explaining what they employ me to do. But in my dreams it would look like this:

HELP WANTED—WRITER TO TRAVEL

Applicant must be willing to visit the farthest, wildest, most strange, and sometimes beautiful places on Earth, and try almost anything once. You will never have to wear a business suit, and Human Resources will not know your identity. Meals are provided and may include giraffe jerky, warthog sausage, and an occasional side of beetle larvae.

Pay will vary. (Translation: Not much, or if possible less.) The employer does not provide medical insurance or any pension. On the other hand, all reasonable expenses will be covered. OK, yes, that will include the mud-walled hotel in western Uganda with one toilet serving all rooms. And, OK, it's not really a toilet, but a hole in the floor. And yes, yes, yes, you may experience near total liquefaction there in the form of the week-long gastrointestinal calamity called giardia. But could we get back to our requirements, please?

Must love animals. The candidate will travel on foot with !Kung San hunters stalking a leopard up a rocky canyon in Namibia while earnestly hoping that the leopard is not, in fact, stalking them—something leopards do much better even than !Kung San hunters.

The candidate will sit down in open country and commune with a pack of African wild dogs. Disregard that rumor about their being vicious man-eaters. They're sweethearts. Honestly.

Willingness to shed conventional norms a requirement. The candidate must be able to contemplate in a nonjudgmental way even the animals that happen at the moment to be having sex, possibly incestuous, on his forehead.

Familiarity with the scientific method a plus. You will at times, for instance, want to fling carefully weighed chicken carcasses into various piranha-infested bodies of water, and time how long it takes to reduce them to bone fragments and pin feathers. Does this feel like a nice day for a swim?

Occasional loneliness is a hazard of the work. The ability to say, "Bartender, I'd like another, please," in as many languages as possible will be an invaluable tool in your chosen profession.

Diplomatic skills a plus. Should you be traveling on the Rwanda border at a time when a local warlord is offering $1,000 for the head of an American, and should your inebriated British travel companions begin loudly singing "The Star Spangled Banner" when you approach, feel free to mutter,

"O! Les cochons Américains!" *while moving rapidly in the opposite direction.*

For our ideal candidate, this job offers moments of sublime beauty most people only get to dream about. Flying low and slow down the Rio Grande in a 50-year-old Piper Cub, for instance, you will have roseate spoonbills on your wingtips, sleek and lovely, in the afternoon sun, as winged flowers. In the llanos of Venezuela, the egrets will cluster in the trees along a marsh causeway like bright white Christmas ornaments, rising up in clouds as you approach and settling down behind as you pass by. You will wake up one morning in the Himalayas to find clouds traveling in a vast numinous river down a high mountain valley and tumbling over a precipice into the abyss, like a waterfall of angels.

You will feel at such moments like the luckiest person on Earth.

Is this a life you would care to pursue? Are you our ideal candidate?

Additional details follow below.

RICHARD CONNIFF
June 30, 2008

Wild Dogs

Somewhere deep in Botswana's Okavango Delta, a million miles from nowhere, a dog named Nomad leads his pack on a wild chase through the bush. The sun paints a gaudy orange stripe across the horizon. Night threatens at any moment to rush down and set the leopards afoot. Our Land Rover bucks and jumps through a dense thicket of mopane trees, and then breaks out onto a flood plain through the catpiss smell of windshield-high sage. Giraffes and tsessebe scatter ahead of us, kicking up panicky clouds of dust. Nomad is the orphan child of a male named Chance and a bitch named Fate, and maybe more sensible men would take the hint and give up, go home, get dinner.

The driver, a wildlife biologist named John "Tico" McNutt, spots a herd of impala, fast food for the wild dogs we are following. But there are no dogs in sight. He listens to his earphones for the signal from Nomad's radio collar, and then the Land Rover dives back into the forest. "Uh-oh," McNutt says, as he muscles the wheel one way and then the other through a thicket of obstacles. "Uh-oh." He circles a tree once to get his bearings, and then lurches off in the direction that makes his earphones ping strong as a heartbeat. Thorny acacia branches howl down the sides of the truck and leap in at the open windows. A rotten log explodes under our tires, showering us with debris. "Captain, we've been hit!" McNutt reports and guns the engine.

And then we see the dogs out ahead of us, long-legged and light-footed, seeming barely to skim the ground as they hunt. They stand about 2 feet tall at the shoulder, mottled all over with patches of yellow, black, and white. Their ears are round as satellite dishes, and their mouths are slightly open. Everything about them as they glide through the mopane seems effortless. Then they vanish, dappled shadows moving among the dappled shadows of dusk.

The common name—African wild dogs—is unfortunate, suggesting house pets gone bad. In fact, *Lycaon pictus*, the lone species in its genus, is utterly wild and only distantly related to our domestic dog or any other canid. Wild dogs most closely resemble wolves in their social behavior, though they are smaller and more gentle. They are like wolves, too, in that humans have vilified and persecuted them into extinction over most of their range.

Only a few decades ago, wild dogs roamed throughout sub-Saharan Africa, in every habitat except jungle or desert. A traveler in the 1960s found them living even in the snows of Mount Kilimanjaro. But they hang on now in just a few isolated pockets, with a total population estimated at fewer than 5,000 animals. They are as endangered as the black rhino, but less celebrated. Farmers still trap them because wild dogs sometimes eat their calves. Hunters occasionally shoot them because they think the dogs steal their game, or because they abhor the dogs' reputedly barbaric killing methods. Until as recently as the mid-1970s, even national park managers routinely killed them. The lore was that wild dogs are an "abomination," capable of killing humans, practicing cannibalism, and whenever possible subjecting their prey to a lingering, brutal death: A pack will chase an animal relentlessly, according to various lurid accounts, "tearing away ribbons of skin or lumps of flesh" until the terrified victim "sinks exhausted, when the pack continues to rend out pieces from the living animal . . ."

One misguided hunter dreamed, in 1914, about the "excellent day . . . when means can be devised . . . for this unnecessary creature's complete extermination." Only now, with that day upon us, has it dawned on people that maybe wild dogs aren't so bad after all. They do indeed kill by disemboweling their prey. But death is typically quick, and no more barbaric than the noble lion using its jaws to strangle a flailing zebra. Wild dogs also run in packs, as alleged. But within the pack they practice family values to a degree that would please, or possibly shame, our leading politicians.

Happy to Be a Bore

The Okavango Delta is a 6,200-square-mile expanse of floodplains and sand ridges, one of the last places in Africa big enough to accommodate wild dogs in their accustomed freedom. Tico McNutt began studying the dogs here in 1989. He is tall and lean, with blue eyes and a second-day stubble. Over the years, McNutt has followed the lives of dozens of wild dog packs and hundreds of individual dogs. He knows almost every dog in his study area by the distinctive mottling of its fur. Often, he knows its parents and grandparents as well, allowing him to construct detailed genealogies and observe the rise and fall of dynasties. He names his packs according to theme, and the names sometimes betray longing for his Seattle roots. There are packs named for weather (Squall, Typhoon, Tempest), movie stars (Dustin, Streep, Uma), and beers (six litters, "a lot of beers").

He and anthropologist Lesley Boggs, his wife, live in a stand of trees next to a dry floodplain on the edge of the Moremi Game Reserve, in the heart of the Okavango. Their camp is improbably settled and homey, with a basketball hoop in the driveway and a kitchen tent softly lit by kerosene lamps. They go to sleep to the hyenas coyly calling "ooo-*WOOO*-ooo" and wake up to the francolins, plump little seed-eating birds, bawling like crows just outside the tent. A hornbill named Hominy lives in camp and routinely steals tidbits from the table.

The dogs McNutt has collared wander through a study area roughly the size of Rhode Island, much of it roadless. He tracks them at times on foot or in a microlight airplane, but mainly by bushwhacking in his Land Rover. He gets three or four thorn-flattened tires a week trying to keep up, and the pink patches on his inner tubes look like polka dots. When his engine overheats, he cleans the debris out of the radiator screen with a feather from a marabou stork.

Driving out from camp one morning, McNutt picks through the dusty gray tangle of hyena, lion, spring hare, and francolin tracks to point out the footprints of wild dogs. "They're very symmetrical, very line-of-direction," he says. "It kind of reflects the balance and light-footedness of the animals as they're moving." He eases down the road, head hanging out the window. "There are three or four dogs here. Cool." He accelerates. "We might just catch up with them."

A few minutes later, he spots a dog moving through the woods. "It's Ditty. She's hunting." Two yearlings join her. Their high bellies testify that they have not eaten, but it's time to knock off for a midday rest. They lie down in the shade nearby, undisturbed by McNutt's familiar truck—until I open the side door and take a seat on the ground. This isn't necessarily a bright idea: Sitting down in a group of wild animals is the sort of dumb trick that gets bush-macho day-trippers ripped to little bits by irritated lions, and quite rightly. Predators deserve more respect than that. But the image of wild dogs as wanton killers is so viscerally embedded in human mythology that it bears first-hand refuting, and McNutt has assured me that I will lose no more than one or two lumps of flesh.

The two yearlings immediately lift their heads. They stand and separate. The sharp edges of their carnassial teeth seem to glint with a scintilla of truth in the old lore. One dog circles behind the truck and begins to creep toward me from the right. The other pads softly through the mopane scrub, head down, and draws closer on my left. A recent comic strip clipping about "Our Fascinating Earth" leaps to mind, characterizing wild dogs as one of "the most vicious of African carnivores . . . and among the few animals that MAY ATTACK MAN." The dogs advance to within 10 feet. Ditty suddenly appears between the other two and strides boldly up to the back of my neck. She sniffs once, then drops back, and all three dogs move off, their curiosity satisfied. They flop down in the shade, having deemed me rather a bore.

"These are wild animals," McNutt says, when I get back into the truck. "They eat animals the size of us all the time. And they're hungry. And yet they showed no aggression whatsoever." In the course of his fieldwork, McNutt has been rammed by an angry hippo, choked with dust when a charging lion skidded to a stop beside his truck, and cornered in the camp shower by a deranged honey badger. But he has never been injured by a wild dog. Ditty's pack has a musical theme, so we name the two yearlings that did not eat me Lyric and Chorus.

Wild Dog Camp

My sons Ben and Jamie were 12 and 14 when I visited Botswana for the first time and they came along to write and take pictures for a blog called *Safari Brothers*. Ben was still a meat-and-potatoes guy then, but at a restaurant on the drive up to Maun, Jamie ordered Phane ka Bogobe, billed as "juicy caterpillars from the mopane tree, full of protein." (Later we had to stop the car so he could vomit on the side of the road.) From Maun, directions out to Wild Dog Camp were not highly detailed. It was something like: Follow the dirt track two hours north and take a left. So we loaded up our battered four-wheel-drive Toyota rental and headed into open country, our wheels sometimes churning hub-deep through the sand and dust, with giraffes drifting past like sailboats. I hadn't packed right, so the bouncing and lurching turned our groceries into a sticky swamp of orange juice, yogurt, and several dozen crushed eggs, which I had to slop out with buckets of water when we eventually arrived.

Wild Dog Camp, where Tico McNutt and Lesley Boggs lived and did their research, was perfect, sheltered within a stand of trees looking out onto the Okavango Delta floodplains. There was a big tent, fenced in on two sides with bamboo, for the kitchen and dining area. The long drop was no rude hole in the ground but a neat hut with a toilet in the middle. The shower, supplied by a black barrel on a platform, stood in the open on the edge of camp. A previous guest had been showering there one time during a drought, Tico said, when he heard a slurping sound from the other side of the bamboo screen. It was a thirsty lion, which soon came around to the shower side, causing the guest to run naked and screaming through camp. This put Ben off the idea of showers. He also wasn't too sure we were joking when we said that, on account of the lost groceries, the meat at dinner would mostly be wildebeest testicles. We pitched *our tent off on our own and went to sleep that night to the sound of lions growling not too far away, a fathomless bellow that deepened and grew louder until my eardrums rang, and then died back down into a sort of bubbling, throaty, airplane-propeller backwash.*

Safari Brothers went up online and quickly became a popular hit. The boys were watching a few days later when I got out of the vehicle and sat down in the open to see if the wild dogs would eat me. I'm not sure if they were feeling the journalistic hunger for good copy (and bloodshed is always nice), or if it was just normal teenage hostility, but I am pretty sure they were rooting for the dogs.

The Twinge of Social Conscience

In truth, what impresses McNutt about the dogs isn't their viciousness, but how gentle and considerate they are with one another. One day we find a pack lying near a great, pyramidal termite mound, nose to rump, like any heap of idle, flea-harried house dogs. But every heap has its etiquette: One of the dogs stands, walks 10 feet away from the others, sits, and claws furiously at his neck with a hind leg. Then he returns to his place in the heap. A social nicety, McNutt theorizes, lest he spread his parasites to a neighbor. Another time when he had just collared a dog and was waiting for it to regain consciousness, a sibling darted in, grabbed the dog by the collar, and dragged it back to the safety of the pack.

Their highly evolved social etiquette also bears on much larger issues. This heap of dogs, for instance, got its start as a pack in the usual fashion when three brothers from a pack named for mountains joined up with two sisters from a pack named for islands. In any pack, only one male and one female do most of

the breeding. The other adults spend their lives helping to rear
nieces and nephews. At the moment, in a burrow underneath the
termite mound, a female named Cypress (for an island in Puget
Sound) is nursing a new litter. She slouches up out of her burrow
and approaches one of the other dogs. Dipping her head down
under his mouth, she makes a soft mewling sound. His belly
begins to heave in response.

By our standards, what follows may sound like an abomination. By theirs, it is selfless everyday caregiving. A nursing
female depends on the other dogs to gorge themselves at the kill
and then regurgitate back at the den. Some dogs will heave up a
portion of their meal seven or eight times a day, especially once
the puppies are weaned and begin to beg. The demands of this
kind of food-sharing are probably the main reason most packs
can support only a single breeding female, and here the social
etiquette can turn harsh. To reduce competition, McNutt says,
the dominant female will often take over or even kill a sister's
litter. These aren't *our* family values, but they are family values
nonetheless. "If it's a small pack, maybe five or seven animals,"
he says, "they're better off having the experienced hunters out
hunting, not back at the den rearing young."

The hunt begins one afternoon with the arrival of a small
procession of ghouls—hooded vultures, a hyena, and us, all waiting for the wild dogs to go out and kill, a chore for which the dogs
themselves appear at first to have no great enthusiasm. They
drag themselves up from the heap and mill around greeting one
another. They lean forward and languidly bridge their back legs
out behind. One of them moseys off. "*This*, believe it or not, is it,"
McNutt says. The others tag along in loose file, with a desultory
wagging of tails.

A subordinate named Blackcomb takes the lead, climbing
up on a termite mound to peer over the grass. The pace picks up

to a trot when there are antelope in sight, and then drops back to a walk when their prey escape. McNutt's Land Rover lurches and zigzags to keep up, and the vultures and the hyena hopscotch behind. The harsh glare of midday softens and the shadows grow longer and more dangerous. The dogs hunt in eerie silence. We hear a single sharp bark—an impala warning its herd—and the dogs instantly jump their pace up to a full run. They are capable of pursuing their prey at 25 miles an hour, with bursts up to 35. But when we catch up with a couple of dogs a few minutes later, they are disoriented, seeming to have lost sight of one another and their prey. Blackcomb is absent, so McNutt keeps our truck bumping cross-country, in response to whatever he is hearing on his earphones. We find Blackcomb before the other dogs do, with his nose in the warm belly of an impala.

He has made this kill by himself, and the victim's supposedly slow, brutal death appears actually to have been instantaneous. The impala lies in a single bright patch of blood in the grass. Blackcomb feeds, looks up, feeds again, and finally leaves to bring his pack-mates to the kill. The feast that follows takes place, like the hunt, in silence. The dogs grip the carcass from opposite sides, and then lift their heads and yank back in unison, as if on the count of three. The only sound is breaking bone and shredding muscle. Cypress, who has come out from the den, twitters softly, and her sister Gabriola backs away.

"The thing that distinguishes wild dogs is that they're so easy-going with each other," McNutt remarks. Where wolves would enforce their hierarchy by snarling and showing their teeth, "you can't help but notice how quiet and cooperative the dogs seem to be in the same contexts." And yet, as Gabriola searches for scraps on the outskirts of the kill, it's clear that a hierarchy is operating here, too. Because the subordinate adults are last in line at the kill, says McNutt, they'll be more motivated to make a kill next time. The risk of leading the hunt brings a subordinate the reward of cramming its belly for a few minutes, like Blackcomb, before the twinge of social conscience causes it to bring in the rest of the pack. "It's a neat system."

That Old Cat-and-Dog Thing

On the way back to camp, McNutt speculates on why wild dogs have evolved into such thoroughly social creatures. They typically travel in packs of about 10 individuals, in part because group living comes at little cost. The most one dog, weighing 40 to 80 pounds, can stuff down its gut at a feeding is about 10 pounds of meat. But their prey average more than 100 pounds. So food for one is food for a crowd. Each extra mouth also brings a pair of those acutely sensitive satellite-dish ears, for added vigilance against "kleptoparasites" like the hyenas, which might easily steal the kill of a solitary dog. The pack also provides protection against the bane of wild dog life—lions.

One evening when Blackcomb is again leading the hunt, he suddenly stops for no visible reason and rears up on his hind legs. His brother Tremblant joins him, peering a hundred yards ahead and making a low, rolling *ru-ru-ru* growl, which means there's a lion out there. The lion yawns massively in the face of the four dogs. "It's that old cat-and-dog thing," McNutt remarks. The lion eventually gets up and plods off into a field of phragmites, the feathery seed heads backlit by the setting sun so they flame like a thousand torches. Blackcomb and the others follow, close enough to nip at the lion's haunches. The lion spins on them and snaps, but continues his retreat. Then two more lions appear and the dogs suddenly recall, with a parting *ru-ru-ru*, that they had an appointment with an impala on the far side of town. In one study, predators—almost always lions—killed 42 percent of all wild dog juveniles and 22 percent of the adults. Humans are the other great cause of wild dog mortality, and these two factors, combined with the footloose behavior of the species, are the reason wild dogs present such a challenge for conservation.

Except during the denning season, wild dogs seldom stay in one place for more than a day or two. In the Okavango a typical pack wanders through a home range of about 175 square miles, and more than four times that in the Serengeti. Few national parks in modern Africa are big enough to sustain a healthy population of wild dogs. And in almost every national park, the dominant species—and the most popular tourist attraction—is the lion. If the dogs seek refuge from lions by going outside the parks, they quickly come into conflict with humans, usually after they kill one of the cattle that have displaced their traditional prey.

Thus until the 1990s many attempts to repopulate parks with wild dogs failed dismally: "Starved or killed by lions within 4 months . . . Shot on nearby farm . . . Left the reserve and were poisoned." The first important exception was at Madikwe Game Reserve, a day's drive from the Okavango, on South Africa's border with Botswana. Madikwe is an experiment, an artificial park of about 290 square miles, created over the past 19 years on derelict ranchland, primarily to bring tourist revenue into South Africa's Northwest Province, and only secondarily for conservation. It's enclosed by a fence 110 miles long, built with steel reinforcing cable and a 7,000-volt electric wire. The consortium that created the park established a balanced population of predators, including just enough lions to gratify tourists. But they chose to make wild dogs a featured attraction. One Madikwe staffer puts it this way: "Lions are common as muck in South Africa. Wild dogs are not."

Madikwe got its wild dog population started in 1995 by putting together a sort of blind date between wild-caught males and captive-bred females. Two packs now coexist there, and their offspring routinely get shipped out as they approach maturity, helping to form new populations at six other nature reserves, game reserves, and national parks around South Africa. To maintain genetic diversity, these parks swap breeding stock according to studbook guidelines, much as zoos do now. What Madikwe promises is a future in modern Africa for wild dogs—if only as a managed, marketed, and fenced-in species.

This is an approach Tico McNutt finds deeply, almost inexpressibly, disturbing. "I don't believe we're going to get very far," he says carefully one evening when we are out watching dogs, "if we justify conservation only by assigning an economic value to an animal or an ecosystem. Surely that's not the only reason. It's not the reason I'm interested in conservation."

Even the creators of Madikwe argue that it would be better to preserve existing wilderness than to attempt to re-create it. But they also say that economic values are what actually motivate people to save a wild area in the first place. "You don't realize its value until it's gone, and then it takes an enormous amount of capital to reestablish it," says Richard Davies, the original project manager for Madikwe. As of 2008, the North West Parks Board has put about 250 million rand ($31.5 million) into the park, and earns about 20 million rand ($2.5 million) in fees a year. Private companies operating 30 tourist lodges within Madikwe have invested another 500 million rand ($62 million). The Parks Board says it does not track how much the lodges are earning, but most lodge employees come from surrounding communities, and two communities own lodges directly. "There's a really strong lesson here for countries to the north of us that are squandering their wildlife," says Davies, and he means Botswana in particular.

Death by Fencing

This is the dispiriting subtext to Tico McNutt's research: He names his wild dogs, records their genealogies, and chronicles their itinerant lives—in the expectation that they will not be able to live this way much longer, even in a wilderness as vast as the Okavango. On the surface, it's a familiar story of villagers with cattle steadily encroaching on wildlife: In Shorobe, a cluster of mud huts on the edge of the Okavango, a group of threadbare farmers sit in the dust to talk, beside a well they have just drilled for a new water hole, and they sound like livestock ranchers everywhere. They long to kill predators. They gripe about government compensation programs, which pay for their lost animals slowly or not at all. They live outside the loose perimeter known as the southern Buffalo Fence, and they routinely lose livestock to roaming lions, hyenas, and wild dogs.

When I suggest it might be better not to keep cattle this close to a wildlife refuge, my translator does me a favor by refusing to translate: "Wildlife belongs to the government, and livestock belongs to the farmer. If it gets into the farmers' minds that you think wildlife is more valuable than cattle, then you will be starting a fire." And he adds: "According to Botswana culture, you cannot live without cattle." It could be Wyoming, outside Yellowstone National Park. But the farmers are also attuned to new possibilities. They envy another village up the road, which operates its district as a wildlife management area, and profits from concessions for hunting, sightseeing, and tourist lodges. "We can live with wildlife on one side of the Buffalo Fence, and livestock on the other," a farmer says. "All we want," says another, "is some benefit from our natural resources."

Environmental critics say the larger threat to the Okavango and its wild dogs is commercial cattle ranching. This industry is heavily subsidized by the European Union, which opened its markets to Botswana beef in 1975 on the condition that the cattle come from disease-free areas. To limit disease, the country built an extensive network of veterinary cordon fences. These fences have cut off ancient animal migration routes nationwide. In one notorious incident in 1983, 50,000 wildebeest piled up dead against a new fence that prevented them from reaching water. Populations of some species have plummeted by more than 80 percent, just since 1978, and the fences have lately begun to close in around the last great enclave of wildlife in the Okavango. With hundreds of miles of new fence being built, range for large mammals and their predators has continued to dwindle.

According to a report from the University of Botswana, the economic benefits of the European subsidy program have gone almost exclusively to commercial ranches controlled by the nation's wealthy ruling elite, not to rural villagers. The same powerful interests are likely to benefit if the Okavango flood plains are converted to ranchland. "People aren't starving in Botswana," Tico McNutt says. "It has to do with a small number of people getting an economic gain out of it."

Four Females from Hell

Out in the Okavango one evening, McNutt and I are talking about dogs and doing our best not to think about all that. Probably we should be savoring all that is sublime and unfettered about wild dogs in their natural element. But the truth is that for the moment we are just having a good time. We joke about the sly twist of destiny that caused three brothers from the Painters pack (Braque, Bacon, and Rothko) to hook up for a time with the Four Females from Hell, before settling down with a trio of females named for single-malt whiskeys (Tamdhu, Islay, and Talisker). It occurs to me that following the different packs as their lives unfold must have a soap-opera feeling for him.

"The thing that motivates me most to find tracks and stay on them all day and go out again the next day," he says, "is to find out if it's one of the hundreds of dogs I've come to know. You've been with their mothers and fathers when they were born, and you see them grow to reproductive age and then disperse and disappear. When you find them again, it's exciting."

"So tell me about the Four Females from Hell," I say. Their names, he says, were Trumpet, Viola, Tympany, and Bell, and at various times he saw them with seven different groups of males, several of which died or disappeared soon afterward. Among the suitors, somewhere between The Painters and Toto from the Wizard of Oz pack, was a male named Piccolo. "Then the Females disappeared," McNutt says, and after two years he figured that lions or farmers had killed them. But one day a new pack showed up in his study area and it dawned on McNutt that the male was Piccolo and the female was Bell. "I found them hanging out together with yearlings," he says. They had become the parents of Ditty, the same dog that sniffed at the back of my neck, and Lyric and Chorus, who did not eat me. "At least one of the Four Females from Hell had successfully reproduced and stabilized," McNutt says, gratified, with an "I knew the bride when she used to rock-and-roll" sort of smile.

Later, a solo male showed up on the fringes of the pack, and it turned out that McNutt knew him, too. His name was Newkie (short for Newcastle), a Beer-pack dog whose older brother had once courted the Four Females from Hell. Newkie also had a history in McNutt's notebooks. McNutt had watched him grow to reproductive age and then strike out on his own. While he was away, an epidemic hit the Beer pack, possibly rabies or canine distemper picked up from a villager's dog. "Newkie came back and found everyone dead," McNutt says. He settled into his old home, finding solace for his social nature in the scent marks of his pack, which lingered like ghosts for months afterward. Then he began to shadow the Music pack, hoping to lure away Ditty, or possibly to replace Piccolo as the dominant male. Ditty showed no interest, and Piccolo repeatedly pushed him off. But McNutt noticed that Piccolo's rebellious son Riff sometimes ran interference for Newkie against Piccolo. Now Riff had actually left his home pack to join up with Newkie. Together, says McNutt, they have a better chance of attracting females than either of them would on his own.

But this is as far as the story goes this evening. McNutt cannot say if Newkie will get a girl, or if Ditty will finally strike out on her own, or if Bell and Piccolo will grow old and dowdy together. He will have to stay tuned for the next episode. We watch the birds known as quelea come rolling into their evening roosts, undulating like swarms of insects above the marsh. A couple of red-necked falcons pick off stray birds to eat for dinner. In the distance, a hippo sounds its sonorous bassoon note.

The Land Rover turns back to camp, and it occurs to me that the lives of the dogs are as messy and tangled as our own. As rich with the tidal coming and going of generations. McNutt is wheeling around trees and stray elephants, muttering, "Uh-oh, uh-oh," and I am thinking about another night when I watched a litter of yearlings playing in the dark. A half dozen of them chased each other in a tight circle around a sage bush, diving into the middle and then shooting out the sides. They jaw-wrestled and played tug-of-war with one another. They made mock charges and danced apart, and then stood with mouths slightly opened, eyes bright, seeming to grin. One paused to catch his breath, as if having called time out. Then he crept up to pounce on a littermate and set the chase going again. All this took place, like almost everything wild dogs do, in silence. The sounds we heard in the darkness were the dry rustling of the bush, the huffing of the dogs' breath, the soft, horselike thumping of their footpads on dry earth. A person passing by 50 feet away might have thought there was nothing much going on out there.

We left them like that, dancing together in the darkness. "You always leave wanting to come back," a friend had told me. With luck, the dogs might still be there if I ever got the chance. They would be lying in their doggy heaps or hunting like dappled shadows at dusk. Better to think about that, I figured, than the other possibility, which is that soon there may be no wild dogs at all.

Life on the Web

One afternoon in the courtyard of a hotel on the outskirts of San José, Costa Rica, a biologist named Bill Eberhard bent over a patch of garden where nothing much seemed to be happening. He held up a double layer of old gym socks stuffed with cornstarch and tapped it softly. *Pa-pa-pa-pa-pa-pa-pa-pa.* A cloud of white dust drifted across the garden and settled on every surface.

Like images taking shape in the darkroom, spiderwebs began to materialize. "They come out of nowhere," Eberhard said. There were five or six of them in an area of about a square foot, most of them classic orb webs, with spokes radiating out from a hub. Each orb was perfect in its way, with a thousand or more delicate intersections, each skewed to its own peculiar catching angle, and each, until the moment the powder settled, nearly invisible. For insects on the wing it must have been like swimming through a bumper-to-bumper sea of fishing nets. Except that some spiders, unlike human fishermen, eat their nets and reweave them up to five times in a day. "These are just the ones they've rebuilt since it stopped raining," said Eberhard.

Spiderwebs are everywhere, and if they happen not to be somewhere at the moment—for instance, in the living room you have just compulsively vacuumed clean—they will almost certainly be there soon: When a spider yearns to travel, it climbs up to a high point and pays out enough thread to catch the breeze and balloon itself skyward, with limbs akimbo for maximum float. These aeronautic maneuvers can take a spider more than 2 miles high and 200 miles cross-country, though a typical trip may last no farther than the next bush. This form of travel is so routine that in one study of a 2½-acre field 1,800 spiders drifted in on their gossamer parachutes each day, like paratroopers over France during the invasion of Normandy. All of them armed and dangerous, at least to insects: A British researcher once calculated that local farmland harbored more than 2 million spiders per acre—and that the insects consumed by spiders each year nationwide would easily outweigh the human inhabitants.

The chief weapon in this endless slaughter is, of course, the spiderweb, and the soldiers are mostly female. Males typically abandon web building when they reach maturity and instead wander around making love, not war. But females need the protein from insect prey to produce eggs, and they weave webs throughout their lives.

In the beginning, roughly 400 million years ago, spiders used their silk mainly to weave a hiding place, possibly with a trip wire out front to detect insects. But then the reclusive and wingless spider suddenly took to the open air. "The reason spiders evolved aerial webs in the first place," says Jonathan Coddington, of the Smithsonian's National Museum of Natural History, "is that insects evolved wings." Tarantulas, trap-door spiders, and some other species still use their silk mainly for shelter, but about a third of the 35,000 known spider species (Coddington estimates there may be 135,000 more) are orb-web weavers, and another third weave sheet webs, cobwebs, and other implements of insect death.

Going Nowhere Slowly

You get a sense of how clever these weapons can be on a short walk with Eberhard, a lean, taciturn field biologist on the faculty at the University of Costa Rica and the Smithsonian Tropical Research Institute. One morning at La Selva Biological Station, in the Atlantic foothills of Costa Rica's central mountain range, Eberhard and I went out wading waist-deep in a rain forest stream, with our faces down close to the surface. Watching spiders means narrowing the scope of your world and moving in millimeters, and Eberhard is a master at this.

He does not hesitate to hum to a spider (spiders can associate particular prey with a specific musical note) or to pick one up with his fingers for detailed examination. ("When squeezed gently on the abdomen," he has written, one spider "produced a strong, somewhat disagreeable odor reminiscent of . . . lampyrid beetles and canned string beans.") He has been bitten just once. I asked if this was because the threat of spider bites is grossly exaggerated or because he is careful about which spiders he grabs. "Both," he said.

We were looking for a spider in the genus *Wendilgarda*, which strings a sort of tightrope across a stream and "glues" its web to running water. After an hour of searching, Eberhard called me over and said, "Here's one," indicating a spider smaller than a freckle, suspended above the water between the drooping leaves of a dieffenbachia. He took out the cornstarch and went *pa-pa-pa-pa-pa-pa-pa-pa*.

"Part of the Zen of this is you find a thing like this and you leave it alone," Eberhard whispered, a subtle attempt to get an outsider to throttle back and see the world in spider time. "You get a little powder on it and figure out which lines are connected to which and which plants are connected, so you can see how to move around it without disturbing it."

The powder revealed 13 separate lines down to the surface, like the leaders on a fisherman's trotline. The riffle of the stream kept the end of each line skating back and forth in search of water striders. After about 15 minutes, irritated by the weight of the powder, the spider began to cut away the old lines and replace them. It descended to the surface on a thread, like a mountain climber on a rappel, and dabbed the water with the silk spinnerets on its hind end. I asked how this brief maneuver can attach a silken thread to running water, something humans cannot achieve with our best superglues. Eberhard suggested that the attachment was more like a sea anchor, a burst of threads held in place by surface tension. The spider climbed halfway back up, then down to the surface again, this time applying a length of a special sticky silk to entangle its victim.

"You have an essentially blind animal with a limited nervous system building a complicated structure in an unpredictable environment," said Eberhard. "The spider makes what for a human would be very complex calculations: 'How big is the open space? How much silk do I have? What attachment points are available?' Spiders are not little automatons making the same thing over and over. They're flexible. And they're not stupidly flexible; they're smart flexible."

Among other accomplishments, Eberhard discovered a species that has reduced its web to a sort of spitball on a thread. It imitates the perfume of a female moth to lure male moths into its neighborhood, and then beans them with the spitball and reels them in. He named the species *dizzydeani*, after the celebrated pitcher who used to do the play-by-play on baseball games when Eberhard was a kid in Arizona.

That evening, with a headlamp on, Eberhard set out for a walk in the rain forest at La Selva. We stopped almost immediately to watch an ant walk into a spider's sticky thread, which instantly yanked it into the air, legs flailing. Eberhard turned and again stopped almost immediately, at a web consisting of a single unsticky strand on which insects roosted like birds on a wire. A green spider lurked nearby, its long, thin abdomen curled up like a tendril. Its technique is to creep with vinelike patience toward an insect using the roost it has provided. It reaches out its front legs and gently, almost imperceptibly, tastes the insect with the hairs on its feet. Then, in a flurry, the spider wraps its prey in silk. Eberhard started to walk and, again, stopped. His headlamp picked out a classic orb web, with the concave shape of a satellite dish.

"Watch this," he said. He touched the web from behind, and it sprang forward. Then, magically, it became concave again. The spider, said Eberhard, uses a spring line running straight back from the hub to winch the web into the cocked position. "Some prey, like mosquitoes, fly very tentatively, with their forelegs out, and as soon as they touch a web, they back off." So this web springs out to follow them. Our walk continued like that, stop-and-start from web to web. After a couple of hours, having journeyed through an entire deadly universe in miniature, we turned back. We had covered all of 50 yards.

"That's how it is with spiders," Eberhard said. "You go nowhere."

Oh, What a Tangled Web

The Zen of going nowhere was beginning to grow on me. One day back home, I was watching a spider spin its astonishing construction between my desk lamp and telephone (it was a slow day), and

I suddenly wanted to become a spider, at least for a little while. I picked up the phone (a cataclysm for the spider) and found a climbing instructor named Stefan Caporale, who agreed to help me build my own orb web, in the corner between two climbing walls at the YMCA in Worcester, Massachusetts. Caporale fitted me out with a climbing harness and Jumar ascenders. I'd never done any rope climbing, but with a slingful of metal carabiner clips over one shoulder and a rope bag in lieu of a silk gland over the other, I felt like *Charlotte's Web* meets *Rambo*.

I was, of course, going to have to cheat, starting from the moment I climbed one wall, tied my first line, and looked across 15 feet of open space to the point where I'd be anchoring the opposite end. A spider bridges this span the same way it makes a parachute, by lifting its hind end and paying a length of silk out onto the breeze. This wasn't going to work for me.

It was cheating just to look. A spider knows what's happening around it largely by touch. It relies on as many as 3,000 vibration sensors, called slit sensilla, most of them on its spindly legs. Eberhard had e-mailed me this thoughtful advice on my web building: "Do it (as much as you can) with your eyes closed."

Having tied my line to a bolt hanger, I climbed back down and climbed up the other wall, where I pulled my spanning line taut. Then I shinnied back out the spanning line, trailing rope behind me. The idea was to leave this rope slack and let the middle of it drop down to become the hub of the web. A spider can do this blindfolded. Then it rappels down from the hub and stretches a spoke to the bottom of the web, keeping the whole thing under tension. Creeping out into midair, 15 feet above the concrete floor, I moved by millimeters. My muscles quivered. Then I began to oscillate, until I was flailing wildly from side to side and spinning sweat in all directions. It took me a half hour to get the first few spokes in place. The average orb-web spider, working at an effortless trot, would already have completed an entire web, with perhaps 30 spokes. Many spiders rush to complete their webs in the last minutes before dawn, to minimize their daylight exposure to predators and also to have everything nice for insect rush hour.

My excuse was that I had to link up all my intersections with carabiners and cumbersome knots. A spider does the same thing with a quick dab of fibrous glue, one of as many as six silklike products it may use to build a single web. The spider applies its silks with a half-dozen spinnerets, each resembling a shower-head with several tiny spigots and each spigot connected to a particular type of silk gland. The silk comes to the spigot as a liquid, a soup of accordion-pleated amino acids. To spin thread, the spider reaches back with a hind leg and yanks silk from the spigot. This shears the silk, so one set of pleats tucks neatly into another, forming hydrogen bonds and making a solid thread. The spigot also controls diameter and flow rate; it will spin thicker thread if you put weights on the spider's back and thinner thread if you put the spider in the zero gravity of outer space.

After four cautious, creeping hours I managed to complete my framework, with a total of just nine spokes. But I had yet to weave the real killing surface of the web, a thread that spirals from the hub out to the edge and crosses the spokes to form a sort of fretwork. A real spider builds two such spirals for every orb web. The first is just a temporary scaffold to stand on. As the spider moves around the web, it eats up the sections of scaffold it no longer needs and recycles the silk. The second, permanent spiral consists of sticky silk. The spiders themselves don't get stuck, according to one theory, because they have an oil coating on their feet. They're also careful to walk on the nonsticky framework of the spokes.

Dangling from my web at the YMCA, I realized that I was never going to cut it as a spider. A real spider's web isn't a big, dumb net like the one I was building. It's a dynamic weapon in an endlessly escalating evolutionary war between spiders and insects. For instance, the silk in the framework of the web has evolved to be almost invisible. Thus a fruit fly approaching at 57 body lengths per second may not detect impending disaster until it is 3 body lengths away. But the fruit fly in turn has evolved optic nerves hardwired directly to its maneuvering wings, so it can still reverse direction and escape before hitting the web.

The sticky silk must be extraordinarily elastic, bulging on impact to as much as four times its resting length. To avoid snapping the insect, trampoline-style, back into the stratosphere, some spiders actually hold the spokes of the web tense as they sit at the hub and then let them go slack when their prey hits.

After five hours I finished my web. It was a little lumpy and off-center, and a real spider would probably have been ashamed, not to say extinct. Scaling up from the spider's body size to my own, I should really have been spinning a web 30 stories high between a couple of neighboring skyscrapers. Still, I had built something unmistakably modeled on a spider web, and I was proud.

For a spider this would be just the beginning. The whole point of all the calculations that a spider has made so far has been to catch insects. And the webs are remarkably productive: On average, spiders eat about 15 percent of their own body weight daily. "To do that," Eberhard said, "you'd have to catch four or five rabbits a day, every day. If you didn't have a gun to shoot 'em, that's an awful lot."

Butt in the Breeze

It's easy to feel sorry for the insects. One day I watched a tiny wasplike creature, long and thin, with a purple metallic body. Through a magnifying glass it looked oddly human. It was stuck to a web by two of its legs and by its long antennae, and it was using its forelegs to groom the antennae in an attempt to free them from the glue. Then it got frustrated and lashed left and right, swiveling frantically around the vertical line of the sticky silk. A little later it got its abdomen stuck on one of the other lines of the web, and its struggles soon ceased. There was something terribly poignant about its plight.

But for the spider, insects are not easy pickings. First, the spider must figure out exactly where an insect has landed, usually by jerking on the spokes of the web to see which one is carrying extra weight. A spider with a frail web may need to get to its victim instantly. On average a fly stays in an orb web for just five seconds, and researchers estimate that up to 80 percent of all insects actually get away. But the spider must be cautious too; if the victim is a bombardier beetle, for instance, it might shoot boiling liquid at the spider.

Depending on the insect, the spider may lunge at its victim and inject paralyzing poison. It may also immobilize its victim with a quick wrap of a special gauzelike silk. It may also stop at a safe distance, size up the situation, then cut its victim loose, and walk away. If the framework is badly damaged, it may even need to build a new web. By now the average human attempting to live the spider way would probably have a heart attack and plummet into the abyss.

But the spider persists—not as the evil killer of popular lore, waiting malevolently for its hapless victim. On the contrary, I was beginning to think that the spider passes its life, as do we all, in hunger, uncertainty, fear, and trembling. Life in the web, as one spider scientist put it, means "hanging your butt in the breeze," where it is liable to be attacked by predatory birds or by huge helicopter-like damselflies. Hiding in plain sight is thus among the spider's chief preoccupations. Eberhard recently uncovered one of the most bizarre threats to spider tranquility in the annals of science, and he generously pointed me to an oil palm plantation where I could see the horror story unfold for myself.

I headed down to Costa Rica's Pacific coast, and in an unwonderful forest where the palm trees stood in orderly rows, the Zen of spider watching began to settle me. I walked slowly, stood still, exhaled, looked around. *Plesiometa argyra* spiders were spinning their perfect orb webs in the undergrowth. *Plesiometa* is a harmless, brightly colored orchard spider, small enough to fit on your thumbnail. When the sun slanted down between the trees, the iridescent webs shimmered in the light.

Then as I watched, a wasp flew up and, in a frenzy of long thin legs, stabbed its stinger into the mouth of a *Plesiometa*. The spider lapsed into stillness. The wasp was a slender creature about three-quarters of an inch long, with black eyes, dark violet wings, and orange legs. She curled her hind end under and jabbed her ovipositor against the paralyzed spider's abdomen, deposited her egg, and then flew off.

Ten or fifteen minutes later the spider woke up and resumed its normal life, unaware that it now carried its own killer. According to Eberhard the wasp larva would hatch in a couple of days and make little holes in the cuticle of the spider to suck its blood. Nearby I found a spider going about its normal business with a fat round wasp larva wrapped like a bumper around the front end of its abdomen. The spider was feeding on prey, oblivious to the larva feeding in turn on its blood. For the first week or two the spider would continue to build its orb webs as normal several times a day. But then suddenly the wasp larva would take control of the spider's mind.

It happens at midnight. Instead of waiting till dawn to make its usual orb with thin, fragile spokes, the spider now goes back and forth up to 40 times on the same few spokes. Somehow, the wasp larva has switched on what Eberhard calls a "subroutine of a subroutine" from the spider's web-building software and switched off everything else. The result is a web of just a few sturdy cables, useless to the spider but perfect for the mature wasp larva, which now needs a place to suspend its cocoon for the final weeks of development. "The spider makes this real strong web," said Eberhard, "and then it just stops and sits in the middle." It is waiting to be killed.

I found one of these webs in the morning. In the middle, hanging below the cables, a wasp larva worked on a dead spider like a half-starved child sucking marrow from old soup bones. By midday the larva looked like a glossy, fat cucumber, and it dropped the withered remains of the spider into the undergrowth. That night the wasp larva would spin its cocoon.

Wandering around the plantation, I found these cocoons everywhere, like little Christmas ornaments in the undergrowth, each securely suspended from its silken cables. I knelt to study one such cocoon. As the late afternoon sun shone through the parchment, I could make out the black eye spots of the wasp and the orange of its legs, which wriggled visibly. The adult wasp would emerge sometime after dawn and seek a mate. If it was female, it would soon find a spider to attack and a web to exploit, starting the whole parasitic cycle over again.

I stood and walked briskly out of the forest. Every now and then it is undoubtedly good to see the world through the eyes of a spider—or rather to close our eyes and feel our way around the spider's strange, silken universe. But at the end of the day I was glad to be merely human.

The Value of a Good Name

People who become zoologists generally do so in the face of sober parental warnings that their earthly rewards will be few and mostly of a dubious nature. And the parents, bless them, are right. I got a hint of what they were worried about not long ago while chatting with a group of specialists when one of them remarked with something like pride, "I've had two dung beetles named after me. And a louse."

It isn't the sort of thing a mother can hang on her wall, is it?

Having a new species named after you is of course a great honor; scientists often characterize it as a form of immortality. But even among biological types, it can also be an occasion for dread. Entomologist May Berenbaum became apprehensive, for instance, after she discovered a new species and passed it on to an expert for classification and naming. "The last thing I need," Berenbaum fretted, "is for a beetle whose distinguishing feature is a proboscis fully half the length of its body to be known as Berenbaum's weevil."

But the gazelles, mountain lions, and other animals people might dream about having for namesakes mostly got handed out long ago. New species nowadays tend to be invertebrates. That means you can still get your name put up in scientific lights, but most likely with the help of strange, small creatures that the unenlightened public regards as vermin. Back in the day, U.S. President Theodore Roosevelt had an elk subspecies named after him, *Cervus elaphus roosevelti*. More recently, President George W. Bush was lucky to attain scientific immortality on the back of a slime mold beetle, *Agathidium bushi*.

Ohno, Ima Hogg

The good news is that, at least for now, there are many such creatures to go around. In the 1980s, Smithsonian entomologist Terry Erwin pioneered a new technique for sampling life in the rainforest canopy, hoisting an insecticide fogger with a net underneath to catch the downfall of slaughtered insects. His samples of ants, beetles, and other treetop fauna boosted the old estimate that there are perhaps 10 million species on Earth to more like 50 million, almost all of them insects in need of names.

This is a lot of names, and zoologists sometimes suffer moments of desperation like parents fretting over what to name the new baby. A Texas couple, Sallie and "Big Jim" Hogg, once named their daughter Ima. Likewise, the inventor and jet manufacturer Bill Lear named a daughter Shanda. On roughly the same principle, Erwin contemplated the 1,500 species ready to be named in the ground beetle genus *Agra*, and dubbed one species *Agra vation* and another *Agra phobia*. A colleague, Arnold Menke, gave a wasp genus a name giddy with the elixir of discovery, *Aha*. But since there were already 8,000 known species in his specialty, the sphecid wasps, Menke also threatened to name his next such discovery in a scientific paper titled "*Ohno*, another new genus of sphecid wasp . . ."

Menke is the recording angel of offbeat zoological nomenclature. Until his retirement, he worked in a fourth-floor office at the National Museum of Natural History (NMNH) in Washington, DC, and kept what appeared to be the business card of a Bulgarian rug merchant on the door:

VYIZDER ZOMENIMOR

ORZIZ ASSIZ

ZANZER R. ORZIZ

Repeat this two or three times. It's actually one of the world's great unsolved zoological and philosophical mysteries: "Vy iz der zo many more horse's azziz . . ." It also suggests the pun-prone, wordplay-addled mind-set of the people behind the bizarre scientific names Menke has collected over the years. Among them, for instance, are a fly named *Phthiria relativitae*, a spider named *Draculoides bramstokeri*, and a pair of musical chiggers in the genus *Trombicula* named *doremi* and *fasolla*.

Three Long Years

Naming new species is of course a serious business, and the rules are complicated enough to fill 338 pages of the *International Code of Zoological Nomenclature*. But they boil down to just about the same system of scientific classification that the Swedish botanist Carolus Linnaeus first devised in the mid-eighteenth century. Every plant or animal must have two names, a genus identifying the group to which it belongs and a species to distinguish it from other members of the same genus. These names are usually constructed on Greek or Latin lines. But Linnaeus wasn't trying to turn the biological world into an instrument of Eurocentric cultural imperialism, or at least that wasn't his primary objective.

He mainly wanted scientists to agree on one universally accepted name for every species to avoid befuddling one another with a vast array of different local names.

His system leaves scientists room to coin names from almost any multicultural phenomenon. For instance, Linnaeus himself once named a shell species *Cypraea isabella*, which sounds lovely enough. But according to entomologist Douglas Yanega, another collector of biological curiosities, the parchment-colored, brown-streaked cowrie was actually named for its supposed resemblance to the soiled linens of the Archduchess Isabella of Austria, who vowed, according to folk legend, not to change her underwear until her father won the siege of Ostend. It took him three years. Modern scientists seem comparatively wholesome when they name wasps for cultural icons—for instance, *Polemistus chewbacca* and *P. yoda*, in honor of two well-known extraterrestrials; a fish, *Zappa confluentus*, for musician Frank Zappa; and a fossil turtle genus *Ninjemys*, to celebrate the Teenage Mutant Ninja Turtles.

Marine biologist Clyde Roper once named a squid genus *Bato-teuthis*, from the Greek *batos*, meaning "thornbush," because the squid has a long, thornlike tail and large, bushy clubs. But this was also the middle of the Batman craze, and if you look closely at the illustrations accompanying Roper's article in a learned oceanographic journal, you may notice a tiny bat of unknown species flying out of the squid's oviduct. Roper protests that he thought the editors would white it out.

"Scientific names have much in common with crossword puzzles," a lepidopterist writes, and if the person coining a name "can mystify his fellow entomologists, he will derive sadistic pleasure in so doing." Hidden meanings are commonplace. For example, the formidable *Brachyanax thelestrephones* translates as "little chief nipple twister." Neal Evenhuis of Hawaii's Bishop Museum, who coined the name, says it has something to do with the shape of this bee fly's antennae. The species name of the fly *Dicrotendipes thanatogratus* comes from the Greek word for "dead" and the Latin word for "grateful" and was coined, in 1987, by an entomological Deadhead.

Other names damn while seeming to praise. *Dinohyus hollandi*, for example, is an extinct species of giant pig named for W. J. Holland, a past director of the Carnegie Museum who made his staff list him as senior author on every paper they produced. The name, loosely translated, means "Holland is a terrible pig." A more broadly misanthropic urge also sometimes seems to take hold of some scientists, as when Menke named a wasp species *Pison eu.* There is also a snail called *Ba humbugi* and a chigger named *Trombicula fujigmo*, the latter to commemorate a slang term once common among American soldiers being sent home at the end of World War II. If you had never heard how soldiers really talk, you could translate the acronym "fujigmo" as "Fie on you, Jack, I got my orders." The paleontologist Robert T. Bakker once expressed a wish to find "a voracious, small-minded predator" and name it after the Internal Revenue Service but sadly has not delivered.

Oh, Kiss Me

Some scientists frown on such foolishness. As Menke stepped out of the elevator one day at NMNH, a colleague remarked that when he was a journal editor he never allowed odd scientific names. "I thought they lent taxonomy to ridicule," he added dourly. In this view, naming species is a sacred discipline, with dutiful taxonomists spending their lives hunched over microscopes, like monks working on illuminated manuscripts. Their solemn business is to distinguish between obscure but almost identical species on the basis of body parts most people have never heard of—the clypeus on a wasp's face, say, or the epandrium on a horsefly's behind.

And then the job is to confer dutifully descriptive names, like *longicornis* (long antennae) or *megacephalus* (big- or, ooh, maybe fat-headed). They publish their results in respectable scientific journals to be duly noted by the world's three or four other experts on bees' knees. And they are expected to love the quiet discipline of their work, these humble taxonomists.

Now and then, though, a wild impulse of delight drives one of them to fling off the fetters of convention. On the brink of his retirement, for instance, an American wasp scientist named Paul Marsh published a paper naming one species *Verae peculya*, another species *Heerz tooya*, and another *Heerz lukenatcha*.

Thoughts of love also sometimes distract a taxonomist, as when the British entomologist George Kirkaldy named several genera of bugs *Peggichisme* (or "Peggy kiss me"), *Polychisme*, *Dolichisme*, and, a little too ardently, *Ochisme*. Carl Heinrich, an American lepidopterist, once told an entire short story in scientific names. In 1923, he called a genus *Gretchena*, after a woman with whom he was presumably involved. Heinrich went on to name species *Gretchena delicatana* (delicate), *dulciana* (sweet), *amatana* (beloved), and *concubitana* (possessed). The tale ended sadly with *Gretchena deludana* (deceived). But maybe this was just wishful thinking, as when fly scientist Neal Evenhuis recently named a fossil bee fly *Carmenelectra shechisme*.

Bugs, Bickering, and Bigamy

Romance in scientific names, as in life, can get complicated. Marc Epstein, a lepidopterist at the Smithsonian, and Pamela Henson, who is the Institution's historian, set out not long ago to study the life of Harrison G. Dyar Jr., a Smithsonian entomologist early in the twentieth century. Dyar is remembered, officially, for having propounded something called "Dyar's law of geometric growth" and for bringing more precise standards to entomological taxonomy. Unofficially, he is celebrated for a life of bugs, bickering, and bigamy. Epstein and Henson first tracked down one of the most celebrated myths in zoological nomenclature, that the quarrelsome Dyar named a species *corpulentis* after an obese rival, and that the rival in turn retaliated by naming a moth species *dyaria*. Sadly, the tale is false: Dyar seems to have done plenty of nasty things to the rival in question, but he never named a species *corpulentis* after him. And the genus name *Dyaria* was coined by a New York banker who simply suffered from a tin ear. He was an amateur lepidopterist and cited Dyar as his "faithful co-labourer and friend," apparently under the impression that he was paying him an honor.

But as Epstein and Henson researched Dyar's life, they began to discuss another moth species that Dyar had named *P. wellesca*. It reminded them of "tales more lurid in nature." For years odd rumors had circulated around the Smithsonian that Dyar had a second family, that mysterious tunnels connected his two homes and that his bigamous deception fell apart when his children from both families discovered at a high school function that they shared a father with a passion for insects.

It turned out that Dyar did indeed own the house next door to the one where he lived with his first wife, Zella, and their children. But he used it mainly for rearing moth and mosquito larvae (and possibly also for getting away from his mother-in-law). His third house, however, was home to his spiritual adviser, Wellesca Pollock, to whom he was also married, apparently under an assumed name. But this house was blocks from Zella's and no tunnels connected the dwellings. According to Epstein and Henson, digging tunnels aimlessly was merely "an eccentric hobby" for Dyar, who seems to have enjoyed a superabundance of energy. Oddly, considering his ability to quarrel with almost everyone else, Dyar seems to have maintained warm relations with both women and their five children. He came up with the name *P. wellesca* in 1900. In 1927, perhaps in the interest of taxonomic and romantic balance, he named a species *zellans* after his first wife, whom he didn't divorce until 1916. Epstein says Dyar also named species for "zillions" of other women along the way.

None of this seems terribly surprising, when you remember that scientists, too, belong to that most ludicrous species, *Homo sapiens*.

Linnaeus himself seems to have understood the human knack for folly. He created the Linnaean system to bring order and harmony to the biological world. But one of the first species he named was *Chaos chaos*.

Lemurs in Love

On a steep slope, hip-deep in bamboo grass, in the heart of the Madagascar rainforest she saved, Patricia Wright is telling a story. "Mother Blue is the oldest animal in this forest," she begins. "She was the queen of group one, and she shared her queendom with what I think was her mother."

The animals she is describing are lemurs, primates like us. They are the unlikely product of one of nature's reckless little experiments: All of them—more than 70 living lemur species— derive from a few individuals washed down from the African mainland into the Indian Ocean at the dawn of primate evolution more than 60 million years ago. The castaways had the good luck to land on Madagascar, an island the size of Texas 250 miles off the southeast coast of Africa. And there they have evolved in wild profusion.

Wright, a late-blooming primatologist from the State University of New York at Stony Brook, has made lemurs her life, tracking bamboo lemurs and sifaka lemurs that live in a handful of social groups in Ranomafana National Park. The story she is telling, to a work party from the volunteer group Earthwatch, is one episode in a running soap opera from 20 years of field research in Madagascar. If her tone evokes a children's story, that may be apt. Wright is a matriarchal figure, with straight auburn hair framing a round face, slightly protuberant eyes under padded eyelids, and a quick, ragged grin. The business of conservation has made her adept at popularizing her lemurs, using all the familiar plotlines of love, sex, marital bickering, wicked stepmothers, and murder.

A sifaka lemur perches at the moment on a branch just over Wright's head. The sifaka, a female, is a handsome creature, a little bigger than a house cat. She has a delicate, foxlike snout and plush black fur, with a white patch around the middle. Her long limbs end in skeletal fingers, curved for gripping branches, and with soft, leathery pads at the tips. When she turns her head, her stark, staring, reddish-orange eyes glow like hot coals. Then she bounds away in a series of leaps, a dancer in perfect partnership with the trees.

Wright first visited the town of Ranomafana basically because she needed a bath. She was looking for the greater bamboo lemur, a species no one had seen in decades. Ranomafana had hot springs—and also a rainforest that was largely intact, a rarity on an island where the vast majority of the forest has been destroyed. In the steep hills outside of town, Wright soon spotted a bamboo lemur and began to track it, the first step in getting skittish wild animals to tolerate human observers. "You have to follow them and follow them and follow them, and they're very good at hiding," she says. "It's kind of fun to try to outwit an animal. When they decide that you're boring, that's when you've won."

The lemurs she was following, it turned out, weren't the ones she'd been looking for. They were an entirely new species, the golden bamboo lemur, which even locals said they had not seen before. (Wright shares credit for the discovery with a German researcher who was also working in the area at the time.) On a

return trip, she also found the greater bamboo lemur she'd originally been looking for.

As Wright was beginning a long-term study in Ranomafana of
both the bamboo lemurs and the sifakas, she came face-to-face
with a timber baron in a Mercedes-Benz bearing a concession to
cut down the entire forest. Wright was a newly minted PhD and
she'd gotten a job on the faculty at Duke University. But she was
20 years out of college and still living the parlous life of an unten-
ured academic, while also raising a daughter. Friends warned her
that letting "this conservation stuff" distract her from research
would hurt her career. "But I couldn't have it on my conscience,"
she says now, "that a species I had discovered went extinct
because I was worried about getting my tenure."

Over the next few years, she pestered the timber baron so
relentlessly that he abandoned the area. She lobbied government
officials to designate Ranomafana as the nation's fourth national
park, which they did in 1991, protecting 108,000 acres, an area
five times the size of Manhattan. She also raised millions of dollars, much of it from the U.S. Agency for International Development, to make the park a reality on the ground. Wright had never
built anything, much less a national park. But flying back and
forth from her teaching duties in the United States, she oversaw
hiring of local villagers, construction of trails, and training of
staff at Ranomafana. She sent out teams to build schools around
the park and to treat diseases such as elephantiasis and round-
worm, which were epidemic in the villages. Her work won her a
MacArthur Foundation "genius" grant, and Stony Brook wooed
her away from Duke with an offer allowing her to spend even
more time in Madagascar.

Along the way, Wright also found time to get to know her
lemurs as individuals, particularly the sifakas in five territo-

rial social groups, and watch as their paths crossed in unexpected ways. At about the time she met Mother Blue, for instance, Wright also encountered a young sifaka in Group Two she named Pale Male. "He was a great animal, very perky," she tells the volunteers. "He would play all the time with his sister, roughhouse around, go to the edges of the territory. And then one day, Pale Male disappeared. A lemur's lost call is a mournful whistle, and his sister gave it all day long. Finally she came down the tree and started giving play faces to my daughter, who was also a juvenile. We think they have no idea who we are. But they know exactly who we are." Pale Male later turned up in sifaka Group Three, where he enjoyed an interlude of lemur bliss with the resident female, Sky Blue Yellow. They had a son called Purple Haze.

"Then one night, as often happens, the fossa struck," says Wright. The myth persists, she tells her audience, that Madagascar, like many smaller islands, is a paradise without predators. In fact, it has eagles, hawks, and the nightmarish fossa (pronounced "foosa"), a nocturnal mongoose with a lean, ravenous body.

At night, lemur groups typically climb high up a tree, out on the branches, to sleep in peace. The fossa has a knack for finding them there. It creeps up the trunk of the tree, its body pressed close to the bark. Then it leaps out into space and catches a lemur by the face or throat with its teeth.

The morning after the fossa struck, Sky Blue was gone. Pale Male, badly battered, soon also disappeared, leaving behind his two-year-old son, Purple Haze, who gave his lost call by day and doubtless spent his lonely nights in terror of the fossa's return. Then one day six months later, long after his presumed death, Pale Male came back bringing a new female into Group Three. "Everybody says lemurs are stupid," says Wright, who was there to witness the reunion with Purple Haze. "But that baby was so excited to see that father, and that father was so excited, and they just groomed and groomed and groomed."

Ranomafana, it turned out, was home to more than a dozen lemur species with an endless supply of behaviors in need of further study. Wright went on to build an independent research station there called Centre ValBio (short for "valorization of bio-diversity"), which now employs more than 80 people and accommodates up to 30 students and researchers at a time.

Wright's success has naturally also attracted critics. Some prominent academics complain that she has not produced enough solid science or trained enough students from Madagascar as full-time scientists, given the funding she has received. (Wright points to more than 300 publications from research at Ranomafana.) Conservationists mutter that, when she hears a good idea, anybody's good idea, she is liable to grab it. "A lot of people are jealous of her and complain about her," says Russ Mittermeier, president of Conservation International, who gave Wright the grant that brought her to Ranomafana in the first place. "But, boy, give me 100 Pat Wrights and we could save a lot of primates."

Housewife Yearns to Become Primatologist

Wright was a Brooklyn social worker when her career as a prima-tologist got its start with a purchase she describes now as "almost like a sin." Before a Jimi Hendrix concert at the Fillmore East in Manhattan, she and her husband were visiting a nearby pet shop. A shipment had just arrived from South America, including a male owl monkey, says Wright, "and I guess I fell in love with that monkey."

Selling wild-caught monkeys is illegal today. But this was 1968, and the monkey, which she named Herbie, took up residence in the apartment where the Wrights also kept a large iguana, a tokay gecko, and a parrot. Monkey and parrot soon developed a mutual loathing. One night, the monkey "made a leap for the parrot, and by the time we got the lights on, he was poised with his mouth open about to bite the back of its neck." The parrot went to live with a Buddhist friend of the Wrights'.

The monkey, on the other hand, took hold of Wright's soul. Wright began to read everything she could find about the genus *Aotus*, nocturnal monkeys native to South and Central America. After a few years, she decided to find a mate for her pet. She took a leave of absence from work and headed to South America for three months with her husband. Since no one wanted Herbie for a houseguest, he had to go, too.

"I thought Herbie would be excited to see his own kind," says Wright, of the female she eventually located in a village on the Amazon. But Herbie regarded the female with an enthusiasm otherwise reserved for the parrot. Wright ended up chasing the two of them around a room to corral them into separate cages. Later, this menagerie moved into a 25-cent-a-day room in Bogotá. "I think the truth is, it was 25 cents an *hour* because it was a bordello. They thought it was hilarious to have this couple with two monkeys."

Back in New York, both Wright and the female owl monkey gave birth a few years later to daughters. Herbie turned into a doting full-time father, returning the infant to its mother only for feeding. Wright stayed home with her own baby, while her husband worked, and dreamed about someday discovering "what makes the world's only nocturnal monkey tick.'" She sent off hapless letters—*Brooklyn housewife yearns to become primatologist*—to Dian Fossey, Jane Goodall, and the National Geographic Society.

Eventually she discovered that Warren Kinzey, an anthropologist at the City University of New York, had done fieldwork on another South American monkey species. Wright prevailed on Kinzey to give her an appointment, and she took copious notes: "Leitz 7 x 35 binoculars, Halliburton case, waterproof field notebook . . ." Then she persuaded a philanthropist from her hometown, Avon, New York, to pay for a research trip to study *Aotus* monkeys in South America.

"Don't go!" said Kinzey, when Wright phoned to say goodbye. An article had just arrived on his desk from a veteran biologist who had been unable to follow *Aotus* at night even with the help of radio collars. "You don't have a radio collar, Mrs. Wright," said Kinzey. "I don't think you should waste your money."

But at least in her own mind, Wright had reason to imagine that she could do better. She'd been spending summers at a family cottage on Cape Cod, following her two monkeys as they wandered at night through the local forest. "It was just fun to see the things they would do in the middle of the night. They loved cicadas, and there was a gypsy moth outbreak one year and they got fat. They saw flying squirrels." So she told Kinzey, "I think I can do it without radio collars, and I've just bought a ticket so I *have* to go."

A few days later, she and her family climbed out of a bush plane in Puerto Bermudez, Peru, where her daughter Amanda, age three, let out a piercing shriek at the sight of a Campa tribesman in face paint and a headdress. Then Wright said, "*Donde está el hotel turista?*" (Where's the tourist hotel?) and everybody laughed.

The local guides were nervous about going out in the rainforest at night to help her hunt for owl monkeys. So Wright headed out alone, leaving behind a Hansel-and-Gretel trail of brightly colored flagging tape. She got lost anyway and began to panic at the thought of fer-de-lance snakes and jaguars. "And then I heard this familiar sound, and it was an owl monkey. And I thought OK, I can't act like I'm scared to death. I'll act like a primatologist. *There are fruits dropping down in four places, so there are probably four monkeys.* And I just started writing anything so I didn't have to think."

Near dawn, she heard peccaries—wild pigs—stampeding toward her through the forest, and she scrambled up a tree for safety. "Then I heard this sound above me, and it was an owl monkey scolding and urinating and defecating and saying, 'What are you doing in my territory?' And by the time he finished this little speech, it was daylight. And then he went into this tree and his wife followed right behind him, and I thought, Oh, my god, that's their sleep tree."

She wrapped the tree with tape, "like a barber's pole," so she could find it again at dusk, and made her way back to camp. Six months later, back in New York, she presented Kinzey with the "first study of the behavior of owl monkeys done in the wild by anybody" and got it published in a leading primatology journal. She also applied to graduate school in anthropology. In her second week back at school, her marriage abruptly ended in divorce.

Polite Society

The mother of all lemurs—the castaway species that somehow found its way to Madagascar—was probably a small, squirrel-like primate akin to such modern-day prosimians as the bush baby in central Africa. Prosimians (a name literally meaning premonkey, now used as a catch-all category for lemurs, lorises, and bush babies) tend to have proportionally smaller brains than their cousins, the monkeys and apes, and they generally rely more on scent than on vision. On Madagascar the prosimians have blossomed into an astonishing variety of forms. There are ring-tailed lemurs, red-bellied lemurs, golden-crowned lemurs, and black-and-white ruffed lemurs—so many different lemurs that Madagascar, with less than half a percent of Earth's land surface, is home to about 15 percent of all primate species.

Among other oddities, the population includes lemurs that eat cyanide-laced bamboo, lemurs that pollinate flowers, lemurs with incisors that grow continually like a rodent's, lemurs that hibernate—unlike any other primate—and lemurs in which only the females seem to hibernate. The smallest living primates are mouse lemurs, able to fit in the palm of a human hand. But an extinct lemur as big as a gorilla roamed the island until about 350 years ago. Lemur species also display every possible social system, from polygyny (one male with multiple female partners) to polyandry (one female with multiple males) to monogamy. And just to keep things interesting, the females are usually in charge. Males acknowledge the female's dominance with subtle acts of deference. They wait till she has finished eating before going into a fruit tree. They step aside when she approaches. They cede her the best spot in the roosting tree at night.

Female dominance remains one of the great unsolved mysteries of lemur behavior. One possible explanation for this role reversal is that food resources, especially fruit trees, are scarce on Madagascar, and highly seasonal. It may be that females need to control the limited food supply because of the nutritional demands of pregnancy and lactation. Big, tough, high-maintenance males would consume too many calories, Wright suggests, and provide little compensatory protection against a flash-in-the-night predator like the fossa. Whatever the explanation, the lemur system of low-key female leadership is a source of deep, playful empathy for Wright.

Dominant females don't usually practice the sort of relentless aggression that occurs in male-dominated species such as baboons, macaques, and chimpanzees. They typically commit only about one aggressive act every other day, and "they do it expeditiously," she says. "They run up and bite or cuff the individual and it's very effective. They don't do a lot of strutting around saying, 'I'm the greatest.'" For every aggressive act, females engage in perhaps 50 bouts of friendly grooming, according to Wright's observations. In fact, grooming is so important to lemurs that it appears almost to have outweighed food gathering in the course of their evolution. Whereas our lower canines and incisors stand upright, for biting and tearing, theirs stick straight out and have evolved into a fine-toothed comb plate, for raking through one another's hair.

Not coincidentally, Wright herself exerts dominance in the benign style of her lemurs. "Zaka," she says one afternoon, taking aside one of her best field workers for a bout of verbal grooming. "I have to tell you about how important you are. When we were looking at all the data from the survey you did, it was very nice, *very nice*." She is also a shrewd consensus-builder. When a young graduate student needs a long-term study site for her PhD, for instance, Wright steers her to a remote village called Mangevo. It has the park's only population of black-and-white ruffed lemurs. But it also has a human population beset by poverty and with a hostile attitude toward the park. Wright makes it clear that the student's job isn't just to learn about lemur biology and behavior. It's also to hire villagers as porters and guides, to put some of the economic benefits of a park in local pockets.

One reason Ranomafana has been such a success, she says, "is that I didn't really know the rules of conservation and what people had done before. What I did was brainstorm with the Malagasy here and with the people in the Department of Water and Forests. It was always a group effort. They had to be a part of it, or it wasn't going to work at all."

Stereotypical Males

Given her close sense of identification with female leadership in lemurs, Wright was shocked recently to learn that the greater bamboo lemurs she rediscovered 20 years ago have a dark secret. "Listen to them!" Wright cries, one morning out on Trail W. "They talk all the time. They crack open bamboo all the time. How in the world could I not have been able to follow them for so many years?" The sound of bark being violently shredded came drifting down from the top of a 50-foot-tall bamboo trunk.

Females spend much of their day chewing through the hard outer surface of giant bamboo stems, till the pieces of stripped bark hang down like broken sticks of dry spaghetti. What the lemurs want is the edible pith, which looks to be about as appetizing as rolled vinyl. It also contains stinging hairs and a small jolt of cyanide. But having adapted to tolerate this poison enables the lemurs to exploit bamboo, an otherwise underutilized resource.

"The female is using her teeth to open these bamboo culms, really working—and the male isn't there," says Wright. "And all of a sudden you hear this big squabbling noise and the male appears just as she opens up the bamboo and he displaces her and takes it from her!" The thought leaves her aghast. *This is unheard of in Madagascar!* Then he moves on and takes away the bamboo from the next female."

At first, Wright and her graduate student Chia Tan thought they were simply seeing bad behavior by one beastly male. Then a new male came in and did the same thing, forcing the researchers to contemplate the bleak possibility that the greater bamboo lemur may be the world's only male-dominated lemur species. Wright and Tan theorize that the females cannot hear anything over the racket of their own chewing. Often, they work just a few feet off the ground, where they are most vulnerable to predators. So they need the male to patrol the perimeter and alert them to danger. But they pay the price at feeding time. "It's beautiful to watch," says Wright, "it's *horrible* to watch."

What Happens to Old Females

In another corner of the park, sifaka Group Three is feeding in a rahiaka tree, and Wright is talking about Mother Blue, the lemur with which she has always felt the deepest sense of identification. During the first decade of Wright's work at Ranomafana, Blue had given birth every other year, the normal pattern for sifakas. She had lost one child to the fossa, and two to other causes. She had also raised two of her offspring to maturity, a good success rate for a lemur. Though they can live 30 years, lemurs produce relatively few offspring and most die young.

Mother Blue, says Wright, was not just a good mother, but also a loving companion to her mate Old Red. "They groomed each other, they sat next to each other, they cared about each other." But Old Red eventually disappeared and in July 1996, says Wright, a new female arrived in Group One. Lemurs are by and large peaceful, but they still display the usual primate fixations on rank and reproductive opportunity. Male interlopers sometimes kill infants to bring their mothers back into mating condition. Female newcomers may also kill babies, to drive a rival mother out of a territory. Soon after the new female appeared, Mother Blue's newborn vanished. Then Mother Blue herself went into exile.

"I arrived a few months later and saw Mother Blue on the border between Group One and Group Two, just sitting there looking depressed," says Wright. "I thought she would die. I thought, *This is what happens to old females. They get taken over by young females and just die.*" It turns her mind to the future.

Despite continuing deforestation elsewhere in Madagascar, satellite photographs indicate that Ranomafana remains intact. Partly because of the success there, Madagascar now has 18 national parks. As a conservationist, Wright aims to capitalize on that success and establish a wildlife corridor stretching 90 miles south from Ranomafana. And as a primatologist, she also still yearns to find out what makes different species tick, much as when she was first learning about *Aotus*.

At the rahiaka tree, for instance, Earthwatch volunteers are keeping track of the lemurs as they feed on a reddish fruit about the size of an acorn. The edible part, a rock-hard seed, is buried in a ball of gluey latex inside a tough, leathery husk. It doesn't seem to discourage the lemurs. One of them hangs languidly off a branch, pulling fruit after fruit into its mouth, which is rimmed white with latex. The sound of seeds being crunched is audible on the ground, where Wright watches with evident satisfaction.

It turns out that Wright was mistaken about Mother Blue. The old female did not simply go into exile and die. Instead, she has moved into Group Three and taken up with Pale Male's son Purple Haze, a decidedly younger male. The two of them have a three-year-old, also feeding in the tree, and a one-year-old roaming nearby. Wright is of course delighted with the way things have worked out. (Now a grandmother, she has also taken up with a younger male; her second husband is a Finnish biologist 18 years her junior.)

Mother Blue, says Wright, is probably 28 years old now and has worn teeth. The Earthwatchers are recording how much she eats and how many bites it takes her to eat it. They're also supposed to collect scat samples containing broken seed remnants, to see how well she digests the stuff. Someone squeamishly points out where the droppings have fallen in the tall grass. Wright wades in. She scrounges around for a moment, and then barehands and bags a couple of fresh pellets for analysis back in the lab.

The sun is shining. Her lemurs are alive and feeding enthusiastically. For Patricia Wright, a primatologist on a hill somewhere in eastern Madagascar, it does not get any better than this. She turns and leads her group uphill, deeper into the forest called Ranomafana.

Oneness with Nature

*I*t is mosquito season again, time for entertaining unwholesome thoughts about nature. Just now, as it
happens, I was reading one of my old journals from a trip
somewhere in South America, when I turned the page. There,
flattened next to the binding, was the dark smudge of a mosquito and, on the opposite page, its Rorschach image in dried
blood, probably my own.

All the unheralded charms of the rain forest came rushing back: the way my clothes were always caked and sodden with
mud, the way the howler monkeys roared their jocund welcome
and flung excrement at my head, the feeling of sliding down wet
clay trails and shuffling cautiously over a wobbly one-log bridge
at midnight in the endless rain, with a dehydration headache
welling up behind my dripping brow. But above all, I recalled the
relief of finally making it back to camp, to sleep and to give suste-
nance to mosquitoes.

For those of you who have not had the pleasure of experienc-
ing this almost sacramental moment of union with Gaia, here
is what it feels like. You drop your clothes in a clay-heavy heap,

leaving your puckered flesh bare just long enough for the mosquitoes to roar in like Sooners at a land rush. This causes you to
dive into your individual cocoon of draped netting, and of course
the mosquitoes follow. You spend the next 10 minutes slapping
and spattering winged droplets of your own blood all over the
netting and the sheets. Then, having killed the last mosquito,
you recollect that you have forgotten to pee, climb back out (after
longing, in vain, for a catheter), and do it all over again.

According to my journal for that night, the sound of slapping
finally died away, and there was a brief period during which the
hostility and bone-weariness of the day succumbed at long last to
peace. Across the way, one of my travel companions looked up dis-
consolately at his white shroud and remarked: "I feel like a pupa."

"I was beginning to think of you as a maggot," the tough guy in
the group replied. (Did I mention that I was traveling with field
researchers who learned their manners largely from insects, and
not social insects either?)

"I've got a tear in my mosquito net," someone else said, as he
started slapping again.

The tough guy immediately began speculating on whether it
would be possible for mosquitoes to drain enough blood to kill a
person and how long it might take. (The tough guy had apparently
learned his manners largely from the botfly, a type of insect that
gets its eggs under your skin, where the developing larvae wriggle
and otherwise annoy you for weeks on end.)

"A mosquito only takes a millionth of a gallon per bite," said
the pupa. He was a decent guy who liked to smooth things over.
"It couldn't happen."

"Brazoria, Texas, 1980," said the tough guy. "There were so
many mosquitoes they killed the cattle. Autopsies said half the
blood in their bodies was missing . . ." He enjoyed trying to keep
us awake at night with entomological horror stories.

"I wish I was home," sobbed the guy with the torn netting.

"Home!" sneered the tough guy. I believe he would have spat to emphasize his contempt, were it not for his own mosquito netting.

Then, as if this were the one thing that might be better, he said, "We could be in the Arctic tundra."

He proceeded to explain that spring in the tundra is so short and sudden that the snowmelt hatches all the dormant mosquito eggs virtually at the same instant. The entire population of mosquitoes then has about 20 minutes to mate, find a victim, get a blood meal, and lay a new batch of eggs before winter sets in again.

Some Canadian researchers once forced themselves to sit still in such a swarm long enough to report that they suffered 9,000 bites a minute.

"Those Canadians know how to have fun," said the pupa.

Suddenly sleep whacked me sideways on the head, and I fell into a dream about all those newly emerged little vampires taking wing in one vast, sky-darkening, bloodsucking swarm, their collective need for a meal so sudden and intense that unlucky, half-frozen victims were literally being sucked to death everywhere I looked.

When I woke in the middle of the night, I saw that the tough guy had flopped sideways against his netting. Mosquitoes were now congregating eagerly in a black knot on the other side, threading their proboscises through the fabric into his rump.

I calculated how much blood he was losing at a millionth of a gallon per bite, and I contemplated waking him up, but didn't. I figured it was what he would have wanted.

To drown out the sound of the mosquitoes' bloodthirsty droning, I put on some earphones and a soothing tape called "Rain Forest Retreat." ("A myriad of beautiful butterflies dance in the shafts of sunlight that filter through a warm, soft mist . . .") I slept the rest of the night like a baby, being careful to keep to the middle of the cot.

The World According to Mark

When writers and photographers work together in the field, murder often hovers eagerly in the wings. One time I shared a boat in the Amazon for a couple of weeks, or perhaps it was a year, with Mark Moffett, who specializes in insect and spider macrophotography for National Geographic. Mark is a genius when it comes to his subject matter, but oblivious to almost everything else. Colleagues sometimes wonder if he was an insect in a former life, or aspires to become one through reincarnation, by mastering the invertebrate way in this life. (I once saw him give a talk in which he went groveling across the stage to his new wife as a way of demonstrating the technique certain male spiders use when begging for sex.)

Mark doesn't much like taking pictures of people, who are at least four legs short of ideal. And he had been struggling throughout the trip with the magazine's request that he take an author-on-assignment photograph of me.

Late one night we were out in the rainforest. We'd discovered a tarantula with a big egg sac, like a leathery, puckered boiling potato. We extracted the spider and the egg sac both from the burrow. Then we sat for more than hour in the rain, with the spider between us, watching to see if she would carry her young back home. Mark was bent forward with his elbows on his knees and his camera trained on the spider. I had my pen and notepad ready under my poncho but was privately pretending to be a gargoyle and watching the rain sluice off the tip of my nose. Ants, beetles, grasshoppers, and other creatures appeared on the tarantula, and on us. Eventually, a mosquito landed on the tarantula. "That mosquito's been eating your blood, Conniff," Mark said, as he focused on its swollen red abdomen. The strobe flashed, and he said, "This is my idea of an author-on-assignment shot."

By the kindness of a mosquito, I had entered the world according to Mark.

Bluebloods

Running out across Pleasant Bay on Cape Cod, watching the schools of bluefish dart and thrash below the surface, or the sanderlings feeding on the mudflats, Jay Harrington used to think his life was almost too good to be true. He made his living catching horseshoe crabs, taking them ashore to extract a soda can's worth of milky blue blood from each one, and putting them back alive.

It was more than just a decent living for a fisherman. Horseshoe-crab blood is the only commercially viable source of an essential medical product known as limulus amebocyte lysate, or LAL. It's what manufacturers use to test the purity of every injectable drug, every vaccine or antibiotic, every kidney dialysis machine that we routinely use on the assumption that they will be safe.

"We were proud of what we were doing," said Harrington. "Every time a rescue vehicle goes down the road, you know that they have drugs in there that were tested with LAL. Anthrax vaccines, smallpox vaccines, all those things are tested with LAL." The only alternative to LAL as a way to keep medical products safe is the sort of test that makes animal lovers blanch: Inject potentially contaminated substances into caged rabbits and see whether it makes them sick.

So Harrington, a mild man with deep-set blue-green eyes and a thin gray beard, was perplexed when the U.S. government shut off his main source of horseshoe crabs. He was baffled when the National Park Service told him that horseshoe crabs are not shellfish but "wildlife" and therefore untouchable. "It's completely bizarre," he said. "It's like living in the Twilight Zone." And he was dismayed when the U.S. Fish and Wildlife Service told him his business was a threat to the sanderlings and other migratory water birds which until then he had regarded as among the chief delights of his way of life. So Harrington did the traditional American thing: He filed a lawsuit.

Three Cents a Tail

Cape Cod itself is only about 20,000 years old, a geologic afterthought left behind by retreating glaciers. But horseshoe crabs were scrabbling around in shallow waters even before the first dinosaurs went plodding past. The earliest fossil horseshoe crabs date back roughly 445 million years. Most people know the species *Limulus polyphemus* today only from the primordial shells that turn up on our beaches, and we tend to dismiss them as "rocks with legs." In our usual fearful way, we act as if the tail contains a stinger. In fact, it's mainly useful for propping the animal up and righting itself when it gets flipped onto its back.

The big helmetlike shell serves as a sort of caisson, beneath which the crab can do whatever it wants without worrying too much about predators. And what horseshoe crabs mostly want is to eat, omnivorously. Jay Harrington once caught one with a partly chewed razor clam sticking out of its mouth like a cigar. To measure their appetite, a biologist put a small horseshoe crab in a tank with 100 half-inch soft-shell clams. After 72 hours, one terrorized clam survived. The other 99 had been chewed up so fine you wouldn't know they'd been there in the first place.

So on Cape Cod, people who knew anything about horseshoe crabs generally did not like them. When Chatham fisherman Nick Nickerson was growing up in the 1950s, he used to catch 400 in a day for pocket money, chopping the tails off and delivering them to the shellfish warden for a bounty of 3 cents a tail. It was "mindless savage amusement," he said, and also a public service, because soft-shell clams are big business in Chatham.

Back then, the only people who had much use for horseshoe crabs were the researchers at the Marine Biological Laboratory in Woods Hole, Massachusetts, where H. Keffer Hartline was using horseshoe-crab eyes to explore the nature of vision. A crab has 10 eyes of varying complexity, including 1 in the tail and 2 on its underside near the mouth. But the main pair protrude high up on the shell like eyebrow windows in the roof of a house, and they are big enough and just complex enough to serve as a near-perfect model for studying vision. Hartline's crab work earned him the Nobel Prize in 1967. One of his students, Bob Barlow, an ophthalmology professor at the State University of New York, has since demonstrated that horseshoe crabs also have the remarkable ability to boost their visual sensitivity a millionfold at night.

Also at the Marine Biological Lab, researcher Frederick Bang was studying blood circulation in horseshoe crabs when he noticed how wounded crabs responded to bacteria. White blood cells called amebocytes rushed to the puncture and formed a cheesy plug that gradually hardened and became part of the shell. It turned out that horseshoe crabs had evolved an extraordinarily sensitive defense against the soup of bacteria in the sea around them. Bang and his successors were able to extract and refine the amebocytes. The result was a product, LAL, that can detect even fragments of dead bacteria.

At the time, it took four or five hours to see whether a test sample injected into rabbits would produce a fever, indicating bacterial contamination. LAL now delivers results in around an hour, at a sensitivity of up to 1 part in 10 trillion. Since the 1970s Bang's discovery has grown into a $50-million-a-year business worldwide.

For more than three decades Jay Harrington has sent most of his horseshoe crabs to a company called Associates of Cape Cod, which started in a Falmouth garage. From May to August roughly 1,200 horseshoe crabs a day pass through the bleeding racks at Associates. When a crab comes out of the barrel, it normally folds itself up, bending at the hinge in its back. A lab worker places the crab in a V-shaped acrylic rack and injects a syringe into the cardiac sinus. The blood that comes frothing out into a 100-milliliter jar is a milky blue, because the molecule that carries the oxygen is hemocyanin, based on copper. (Our blood runs red because it uses hemoglobin, an iron-based molecule.) The bleeding takes 30 percent of the crab's total blood volume and lasts about five minutes. Harrington says his crabs are typically out of the water for only about 30 hours, and most sources put the mortality rate at around 10 percent. He sometimes catches the same crabs, with visible bleeding scars, two or three years in a row (their life span is up to 30 years).

From the relatively primitive bleeding room, the horseshoe crab's blood goes on to an increasingly technological round of refinements. But the final product is remarkably simple. A lab technician at Associates demonstrated with a rack of pinky-thin vials, each containing a few drops of LAL and an equal amount of a sample being tested for contamination. After 30 minutes she pulled one or two vials at a time from the rack and turned them upside down. The vials containing pure samples were still liquid.

But the contaminated samples were now trapped by the horse-shoe crab's primordial immune system as an opalescent clot at the bottom of each vial.

A Little Old Lady in Chatham

Harrington's problems started in the 1980s, when the international market for eels and conch heated up. Atlantic Coast fishermen turned increasingly to horseshoe crabs for bait, in part because they are so easy to catch. In May and June the horseshoe crabs gather in the shallows to mate. Fishermen reach down and pick up love-swept horseshoe crabs by the boatload. "If you can see 'em, you can get 'em," a conch fisherman in Hyannis told me. "I don't think it'll ever be an Olympic event."

Delaware Bay in Cape May County, New Jersey, has the largest population of spawning horseshoe crabs in the world, and the spawning there provides an essential food source for migratory water birds. Having flown in without eating all the way from South America, a single bird may gulp down 135,000 horseshoe-crab eggs before pushing on to its nesting grounds in the Arctic. So when people began to think the bait trade in Delaware Bay might be destroying the horseshoe-crab population, it became, in the words of one biologist, "an environmental rallying point" equivalent to the failing salmon runs of the Pacific Northwest. People up and down the coast began to look twice at anybody having anything to do with horseshoe crabs.

Then one day in 1999, a couple of conch fishermen from Hyannis pulled into Chatham after a night collecting horseshoe crabs in Stage Harbor. Exhausted, they left their skiff sitting on the tidal flats, with a full load of horseshoe crabs scrabbling and baking in the midday sun. Chatham has a thriving community of environmentally aware retirees from out of town, and they were outraged. The incident made the newspapers and led to the founding of a horseshoe-crab conservation committee. "This all came about," another Hyannis conch fisherman told me, "because a friend of mine pissed off a little old lady in Chatham. And if that ain't Cape Cod, I don't know what is."

What you think about the plight of horseshoe crabs depends largely on where you're coming from. "It's the bottom," a Massachusetts Audubon staffer declared. "We're fighting about horseshoe crabs. We've taken it all." And a Chatham fisherman said, "You stay right here. I'll go put on my wet suit and I'll be back in one hour and a half with 500 crabs. I'll bet you $100."

"There really aren't a lot of accurate numbers kept," the Audubon staffer admitted, "but they know the numbers are declining. They know that." In fact, census data on horseshoe crabs were almost nonexistent at the time. Moreover, horseshoe-crab populations are highly localized and often isolated from one another. So while the decline in horseshoe crabs in the Delaware Bay was unmistakable, that didn't necessarily mean much for horseshoe crabs in New England.

On the Cape itself, the evidence was inconclusive. Ophthalmology researcher Bob Barlow, who has been keeping track of the crabs at Bourne for more than 20 years, said he was seeing an 80 percent drop in population there and a 95 percent decrease in spawning, apparently because of overharvesting for the bait trade. But no one had attempted a comprehensive study of horseshoe-crab populations on the Cape.

When Lawyers Do Taxonomy

In any case, the decision to close Jay Harrington's main horseshoe-crabbing grounds at the Cape Cod National Seashore and at Monomoy National Wildlife Refuge had nothing to do with evidence.

The closure wasn't even a matter of right or wrong, said Nancy Finley, chief of natural resources at the Seashore. "It had to do with what the law said."

And the law was the point on which Harrington suddenly found himself stepping through the looking glass. To begin with, he was dealing with two very different facilities and two separate federal agencies (though both fall within the U.S. Department of the Interior). The Cape Cod National Seashore is a huge swath of beaches, dunes, and estuaries running up the forearm of the Cape. The National Park Service manages it, with a dual mandate to provide recreation and protect resources, including the "distinctive patterns of human activity and ambience that characterize the outer Cape," a phrase commonly interpreted to include fishing and shellfishing. Monomoy, on the other hand, is a National Wildlife Refuge. It's home to endangered piping plovers and roseate terns, and the U.S. Fish and Wildlife Service runs it with a much narrower focus on protecting wildlife.

Both facilities have traditionally allowed shellfishing in their waters. So when the National Seashore began to worry about getting caught up in bait-trade decimation cascading up from Delaware, the defensive strategy it seized on was to declare that horseshoe crabs were no longer shellfish. They were really "wildlife" and thus off-limits.

"The only thing biological that affected this decision was taxonomy," Finley said. Horseshoe crabs are not shellfish, federal lawyers declared, because they belong to the phylum Arthropoda and are, in fact, far more closely related to spiders and scorpions than to crabs. This sounded plausible, at least at first.

It may even have looked like good science, to a lawyer. But shellfishermen routinely harvest species from wildly different animal groups. Clams are mollusks. Sea urchins are echinoderms. Conch are gastropods. All of them fit the dictionary definition of shellfish: marine invertebrates with shells. If you disqualify horseshoe crabs because they are arthropods, that means lobsters and crabs, both also members of the phylum Arthropoda, cannot be shellfish either. To which Finley replied, exactly right. According to Park Service taxonomy, lobsters and crabs are fish.

A judge in Harrington's lawsuit against the U.S. Department of the Interior eventually concluded that the National Park Service had categorized horseshoe crabs as wildlife only "in order to assert jurisdiction over them," because local governments control shellfishing. But the judge also allowed the ban on horseshoe-crabbing at the National Seashore to stand.

At Monomoy, meanwhile, Harrington ran into a Catch-22. The Atlantic States Marine Fisheries Commission (ASMFC) required him to drill a hole in the shells of crabs he was releasing and attach a numbered tag for a recapture survey. But U.S. Fish and Wildlife Service officials told Harrington that releasing tagged crabs at the refuge constituted a violation of his special-use permit and of federal law—even though the Fish and Wildlife Service was itself helping to manage that study and had the job of collecting the tagged crabs for ASMFC.

So was there legitimate reason to worry about the horseshoe crabs at Monomoy? A videotape made by the Audubon Society suggested that water birds feed on horseshoe-crab eggs and larvae all summer long. On the tape, sanderlings, sandpipers, short-billed dowitchers, and Hudsonian godwits all busily worked the tidal flats. Their bills jittered up and down in the sand like sewing machines. Researchers washed out the crops of a sample group and found that about one-third contained horseshoe-crab eggs.

No one bothered to determine whether horseshoe-crab eggs made up a substantial portion of the diet for such birds, or whether other foods were more important. And even if it made sense to ban the bait trade as a way to maximize production of eggs, no one bothered to investigate whether it made sense to throw out the catch-and-release biomedical use of horseshoe crabs at the same time. But Monomoy officials concluded that harvesting horseshoe crabs in its waters was incompatible with its primary mission as a refuge.

That left people in Chatham wondering if federal officials might eventually extend its ban to all forms of shellfishing. In recent years, commercial shellfishing on Monomoy has employed several hundred people. The prospect of shutting down this business caused one Chatham fisherman to roll his eyes and say, "Won't that be a firecracker?"

Keeping Their Feet Wet

There are of course other places to catch horseshoe crabs. Associates of Cape Cod said it could probably get its crabs from elsewhere in Massachusetts or from Rhode Island. In theory, it might even become possible at some point to handle no crabs at all, if genetic engineering were to permit the manufacture of certain components of LAL without using horseshoe crabs. Though it has so far proved elusive, a synthetic LAL would spare the biomedical industry from oil spills, overfishing, and other vagaries of the seagoing life. They could leave the horseshoe crabs alone to scrabble in the shallows as they were doing back in dinosaur days.

For Cape Cod, on the other hand, the idea seemed somehow sad. "Most people want people to be able to fish here," a Wellfleet fisherwoman told me. "It's so linked up with their sense of place. People respect the people who have managed to keep their feet wet in these crazy times."

When I last talked to him, Harrington's feet were dry. He was working as a carpenter to get through the winter. His lawsuit had fizzled out for lack of money, and the ban at Monomoy and the National Seashore had survived. The bait trade for horseshoe crabs had also been banned in Pleasant Bay. But a compromise was keeping Harrington in business, allowing him and two other permit holders to gather horseshoe crabs for the biomedical trade from the parts of Pleasant Bay outside federal jurisdiction.

Harrington struck me as very much a Cape Cod personality—quiet, balky, and stubborn. For him, there was no better way to spend the summer than catching horseshoe crabs and putting them back again, no better place to do it than Pleasant Bay, out in the sun with the striped bass and the short-billed dowitchers. It was in his blood.

"The only place I've ever fished," he said, in a tone of puzzlement at the curious turn his life had taken, "is right here on the elbow of the Cape."

The King of Pain

Justin O. Schmidt does not stop talking from the moment I walk through the door of his one-story brick house on the outskirts of Tucson. In the living room, a spindly vine grows to the skylight, and he is soon explaining how he got the seeds from Trinidad and induced them to germinate by chewing on them, thus replicating the effect of a bird's digestive tract. Right next to it, an *Amorphophallus* grows. It gets its name (Greek for "misshapen penis") from its exuberantly spiky flower. Unfortunately, his wife puts it outdoors when it blooms. She says it smells like a dead rat.

But let's get to the point: There are terrariums full of venomous creatures lining every wall of the house, and the only condiment on the kitchen table is a tube of Itch-X. When Schmidt's wife phones in, he tells her that a reporter is visiting and asks if she would like to get stung by a harvester ant and "give us a vivid live description of the pain." She declines.

Schmidt has spent much of his career as an entomologist for the U.S. Department of Agriculture and he looks the part—slightly built, with pale blue eyes, and a bank of thick red hair over his forehead. His blue T-shirt is decorated with a *Polybia occidentalis* wasp. He is co-author of the standard text in the insect-sting field, *Insect Defenses: Adaptive Mechanisms and Strategies of Prey and Predators*. But he is better known as the creator of the Justin Schmidt Pain Index, a connoisseur's guide to just how bad the *ouch* is, on a scale of one ("a tiny spark") to four ("absolutely debilitating").

Among connoisseurs of insect stings, it is the equivalent of Robert Parker's wine ratings, and Schmidt doesn't object to the comparison. In faux-Parker mode, he once described a bald-faced hornet sting as "Rich, hearty, slightly crunchy. Similar to getting your hand mashed in a revolving door." Other researchers tend to regard his work with fascination. But they also keep a safe distance.

Schmidt's index is based on ample experience, acquired mostly by accident. Getting stung deliberately, he says, would be like going to the doctor for a shot—too awful to think about. It's also artificial: He worries that the insect might not inject a normal dose of venom. But since his research on bees, wasps, and ants often requires him to hunt down and collect obscure species, he has plenty of opportunity for instructive mistakes.

"What happens is that you've been looking for a species maybe for years," he says. "You finally find a nest and by god you're going to get every one of them. You get your buckets and your aspirators and you start digging away." In the excitement, a few stings are almost inevitable. "So I pay a little attention to the type of pain it is, how long it lasts, how intense it gets."

One morning not long ago, for instance, he was making his way up the winding road to the Monteverde cloud forest in Costa Rica when he spotted *Parachartergus fraternus*, social wasps known both for the sculpted architecture of their hive and for the ferocity with which they defend it. This hive was 10 feet up a tree, and the tree angled out from an eroded bank over a gorge. Schmidt got out a plastic garbage bag and promptly shinnied up to bag the hive.

"There's always a few that get out," he says. So he had taken the precaution of putting on his beekeeper's veil. Undeterred, the angry wasps charged at his face, scootched their hind ends under in midair, and, from a range of 4 inches, squirted venom through the veil straight into his eyes. "So there I was 10 feet up a tree, holding a bag of live wasps in one hand, basically blinded with pain." He slid down the tree like Wile E. Coyote after a tête-à-tête with Road Runner.

But he held onto the nest. In truth, the one thing wrong with the story, at least in Schmidt's distinctly odd retrospect, is that he only got sprayed by the wasps, not stung. He has been more successful on other outings, sampling the stings of about 150 different insect species on six continents. (Antarctica, with no stinging insects, is hardly worth the trip.)

"The U.S. Department of Agriculture doesn't actually pay you to get stung?" I inquire.

"Oh, heavens no. I don't do any of this on official time." In fact, USDA officials prefer that, when he's talking about sting pain, Schmidt play down his government work and play up his affiliations with the Southwestern Biological Institute and the University of Arizona. No point in causing angry taxpayers undue concern. "This is the kind of thing you do on weekends," he adds.

Schmidt doesn't remember the first time he got stung. He remembers instead his first "experience with stinging." In second grade, he plucked a honeybee off a dandelion and applied it to his teacher's arm "to see if it would sting." It did. Later, he took up beekeeping and started getting zapped routinely. But Schmidt didn't begin to think seriously about stings until one day when he was out doing fieldwork as a graduate student, and a harvester ant nailed him between the toes.

"This thing grabbed my attention," he recalls. "For half an hour, it felt like somebody was putting a knife in and twisting it." The area around the sting began to sweat, and the hairs on his foot stood on end. It dawned on him that he had found an excellent topic for a toxicology paper. He set about collecting venom from harvester ants and injecting it into mice. Using a then-standard technique called the LD (for "lethal dose") 50 test, he measured how much of the poison it took to kill half the mice in the sample. "It blew the top off the scale," he says. "This stuff was about 10 times more toxic than any known insect venom." Though a harvester ant can't match a king cobra for sheer venom quantity, its venom is deadlier ounce for ounce. He now calls it "the world's most lethal arthropod venom."

These are the same ants that the Apaches used on their prisoners, at least in the movies. So I ask Schmidt what would happen if somebody were stripped naked, tied down, smeared with honey, and left for the harvester ants.

"They'd probably die of exposure. It would take a huge number of harvester ants to kill an adult." About 890 of them, to be precise, a number Schmidt has extrapolated from his LD 50 results and not from direct human testing. "But these ants are not that terribly aggressive. Not like fire ants. You really have to work at getting them mad enough to sting you."

Curiously, considering American liability law, harvester ants are also what American parents give the kids to play with every time they buy an ant farm. These ants happen to be ideally suited for life in a plastic box: They lack sticky substances on their feet and can't climb out. They also dig like crazy during daylight hours and they're easy to see.

But before I can ask how many harvester ants it would take to kill a small child, another phone call comes in, and I wander off to soak up the ambience. In the murderers' row along one wall of Schmidt's office, some velvet ants look plush and almost huggable, though they are in fact not ants at all but a wickedly painful kind of wasp. The next terrarium over houses a tarantula twitching its fangs in the direction of a 2-inch-long beetle. A little further along, there's a centipede capable of eating Spanish fly beetles, which are otherwise poisonous enough to kill a horse, and, nearby, a horned lizard that eats harvester ants. This battery of creatures serves Schmidt mainly as a bioassay to determine if a newcomer's defenses are any good. It's a bit like going to prison: Unless you have the wrong stuff, and lots of it, the other inmates eat you alive.

When he returns, the conversation shifts to how different species use their venom in the fight for survival. For example, the wasps known in the American Southwest as "tarantula hawks" (members of the *Pepsis* and *Hemipepsis* genera) depend on venom for their macabre reproductive strategy. The female hunts down a tarantula, injects a paralyzing venom into its abdomen, and then buries the spider with a single egg deposited on its back. When the wasp larva hatches several days later, it eats the paralyzed tarantula alive. (But if you pick off the larva before then, the tarantula wakes up and goes about its former life in apparent ignorance.)

The same tarantula hawk's venom has a totally different effect on humans. "If you get stung by one," Schmidt says, "you might as well lie down and just scream. The good news is that by three minutes it's gone. If you really use your imagination you can get it to last five minutes. But that's it, you get on with your life."

Clearly, Schmidt has given more thought to the nature of stinging than is entirely healthy. But he says there is a point to the subject for us all: Venom can be of defensive value to an insect in two different ways. "One is that it can actually do serious damage, to kill the target or make it impaired. The other is the whammy, the pain." Scientists already knew how to quantify toxicity. But they had no way to measure pain, other than through direct experience. The Justin Schmidt Pain Index came into being, he says, as a tool for interpreting an insect's overall defensive strategy.

The Mother of All Stings

Only female insects sting, and sex—or, rather, reproduction—is the reason they first learned how. The whole nasty business got started back in the Jurassic period. Like tarantula hawks today, early parasitic wasp females had a pointy extension on the end of the abdomen, called the ovipositor, and used it to lay their eggs on living caterpillars, beetle grubs, and other hapless victims, usually at a rate of one egg per victim. In time, some lucky Mama Wasp evolved a serrated edge on the ovipositor to saw through flesh and deposit the egg *inside* the body. Instead of trying to hang on as they sucked their living hosts dry, her babies could develop within the shelter of their victims' bodies, until they were big enough to burst forth, *Alien*-fashion, and fly away.

When wasps came visiting, victims naturally put up a fight. But at some point in the primordial struggle, a genetic mutation gave the saw-blade lubricants or other fluids in the ovipositor of some wasp species the ability to paralyze a victim. This made life infinitely less hazardous for the wasps, and the mutation proliferated in subsequent generations. From this eureka moment, ovipositors increasingly adapted to function as stingers, and venoms evolved to thousands of devious purposes having to do with damage, pain, or some unhappy combination of the two. At least 60,000 different species in the order Hymenoptera, including wasps, ants, and bees, now possess some form of stinger. Impression fossils of a wasp from Russia show that this evolutionary flowering was already well underway more than 120 million years ago.

Think of it this way: Our bodies are like software, and sting-ing animals have evolved to be the hackers, the dweebs and mis-fits that have managed not only to break into the system but also to crack the biochemical code. With a little venom, they can pen-etrate cell membranes, manipulate neurons, convert systems of self-defense into instruments of self-destruction, alter the func-tion of the heart, and even, in some extreme cases, cause death.

Yet the vast majority of stinging insects use their venom pri-marily to parasitize tomato hornworms, cabbage loopers, and the like. Insect stinging is thus more a blessing on humanity than a curse: If female parasitic wasps were not out there busily killing agricultural pests, we would starve.

But this is all too easy to forget in a moment of pain. For us, stinging often means nasty encounters with yellow jackets, fire ants, and other social insects that have retained no trace of the parasitic lifestyle. They now sting purely to defend the hive, and they are dismayingly good at what they do. On Justin Schmidt's Pain Index, honeybees rate only a two ("like a match head that flips off and burns on your skin"). But no instrument of biologi-cal terror is quite so thoroughly understood by science.

The Biochemistry of Ouch

When a honeybee stings you, the barbs at the end of the stinger catch in the flesh, and the hind end of the bee rips off. This kills the bee. But what's left embedded in the victim's skin by this sui-cidal sacrifice is a hypodermic syringe, and it is capable of inject-ing venom for about 20 minutes. Contrary to myth, pinching out the stinger doesn't cause it to inject more venom. So don't waste time looking for a credit card to scrape it out; just get the damned thing out as quickly as possible. The amount of venom injected depends entirely on how long it's stuck in your skin. In addition, the barb contains an alarm pheromone, which calls in other bees and incites them to sting, too.

The sharp end of the hypodermic syringe consists of a grooved tube, or stylet, flanked by two sharp knives, called lancets. Each lancet is serrated with seven or more barbs. The ruptured bulb on top of the needle contains a neural ganglion, which causes the lancets to slice up and down in alternation, so the barbs saw their way deeper into the skin. The top part of the syringe also contains the venom sac, and a valve-and-piston arrangement to pump venom down the stylet and into the wound.

A single honeybee contains relatively little venom—about two hundred-thousandths of an ounce of clear, colorless liquid—and injects even less. According to Schmidt, you'd need about 48,000 bees—a whole hive's worth—to get just 1 ounce of venom. Yet even the minuscule dose injected by a single bee can be shockingly effective at driving a predator away from the hive. The venom contains at least 40 different ingredients geared for painful cellular warfare.

Though they employ strikingly different weapons to get there, all venoms have the same basic target: The cell membrane, a two-layer wrapping around all biological cells. It consists largely of proteins and fatty phospholipids. The phospholipids are ingenious molecules with a bulbous head that's attracted to water and a fatty-acid tail that can't stand the stuff. So the phospholipids of the outer layer naturally line up side-by-side, their heads all pointed out to bask in the great liquid sea of life. The phospholipids of the inner layer line up the other way, with their tails to the tails of the outer layer and their heads facing the calm inner sea of the cell. This makes for a stable two-layer membrane, with a water-resistant fatty-acid middle. It keeps the cell as snug as a house that's wrapped in Tyvek and R-19 insulation.

Then the honeybee's venom bursts in, bent on havoc. A peptide called mellitin strikes the opening blow, shouldering in among the closely packed phospholipids of the cell membrane. This throws open the door to a powerful enzyme in the venom, phospholipase A, which rushes in and severs the connection between the head and tail of the phospholipids. The membrane begins to break apart. If the victim of this attack is a red blood cell, hemoglobin spills out in a widening stream until the entire cell dissolves. If the victim is a neuron, damage to the membrane upsets the delicate relationship between the ions inside and outside the cell, causing the neuron to fire little jolts of pain, called action potentials, over and over.

Norepinephrine and other substances in the venom shut off the flow of blood, turning the skin white and keeping the venom concentrated around the sting. Thus the repeated, stabbing pain may persist for five minutes or longer, until the mellitin is gradually diluted and carried away from the area.

At the same time, other substances in the venom are working to spread the pain. A substance called hyaluronidase liquefies the mucuslike glue of the connective tissue, enabling the mellitin and phospholipase A to scramble onto new targets. This so-called spreading factor is common in snake and spider venoms and also, oddly, in mammalian sperm, where it helps clear the path to the egg.

Swelling and redness start to appear because of the mayhem induced by a peptide called MCDP (mast cell degranulating peptide), which targets the mast cells in the skin. These cells are our frontline security system, specialized defensive cells present everywhere the body comes into contact with the outside world. MCDP triggers the mast cells to release histamine and other substances. This dilates the blood vessels—which is a good thing, at least in theory, because it brings macrophages and other immune-system tools to the scene.

But in some sting victims, the mast cells are studded with antibodies specifically attuned to components in bee venom. The tiniest dose of venom unleashes a hypersensitive flood of histamine, which can cause swelling, bronchial spasms, and plummeting blood pressure. Without proper treatment, allergic shock can lead to death in less than an hour, particularly if the victim is old and suffers underlying medical problems. According to Schmidt, such occurrences are rare: about 40 people die each year in the United States from insect stings, and the death rate worldwide also appears to be small. "More people get killed in a year by stubbing their toe on the sidewalk, falling into the road, and getting run over by a truck," says Schmidt.

The Pain Index

Most insect stings, in fact, do no damage at all. They just scare the wits out of us. And this is why they fascinate Justin Schmidt: We typically outweigh any insect tormentor by a million times or more. We can usually outthink it. "And yet it wins," says Schmidt, "and the evidence that it has won is that people flap their arms, run around screaming, and do all kinds of carrying on." It wins because we generally heed the insect's message, which is: "Leave my nest alone."

Stinging, says Schmidt, is a far more complex and paradoxical business than we might think. Insects generally inject far too little venom to do serious harm, even when they attack en masse. So-called killer bees make headlines because they attack so quickly and persistently that they will sometimes chase a victim and inflict hundreds of stings. But since these hybrid bees arrived in the southwestern states in 1990, Schmidt has documented only 14 deaths from envenomation. Dog bites kill more Americans in a single year.

In any case, pain matters more than killing power for the insect's survival. "How does the insect win?" says Schmidt. "By making us hurt far more than any animal that size ought to be able to do. It deceives us into thinking serious damage is being done." A trapped or threatened insect is a bit like Reese Witherspoon in some Hollywood caper, cornered in the bad guy's corporate headquarters. She can't do any real harm. But she *can* hold a lighted match up to a fire detector. Much as the fire detector thinks "massive conflagration" and screams its head off, sting pain fools us into thinking "serious trouble." We panic, and our insect tormentor makes her getaway—or, better yet, stings again. When the pain of a sting is like "turning a screw" in the flesh or like "pulling out tendons and muscles," we tend to steer clear of those insects thereafter.

A reputation for inflicting pain thus liberates stinging insects from their predators and enables them to open up whole new ecological niches. Honeybees can visit flowers by day and not get eaten by birds. The tarantula hawk, which is the most painful stinging insect in the United States, freely roams the deserts of the Southwest, and all potential predators busily look the other way. Many insects adopt bright yellows, whites, reds, and blacks to advertise how painful they can be. Some cheaters even mimic the bright colors or threatening behaviors, though they in fact possess no venom. Male bees, for instance, often curl up as if to sting, fooling even experienced beekeepers into momentary panic.

Pain alone (much less the mere appearance of being painful) isn't enough to discourage *all* potential threats. Animals aren't dumb, and if they keep hearing a fire alarm where there's no fire—no real damage—they eventually calm down and figure it out. Bears learn to put up with bee stings as a cost of getting honey from a hive. Capuchin monkeys will gobble down a wasp nest full of juicy larvae as if it were a ham sandwich—the stings seemingly no worse than a little hot mustard on the side. Harvester ant stings have evolved a whammy wicked enough to scare off toads, amphibians, and almost every other living thing. But the horned lizard has called the ant's bluff by becoming resistant: It licks these ants up with impunity. So while pain is obviously an important line of defense, Schmidt believes the tendency over time is to supplement mere pain with "truth in advertising"— that is, to make the pain more persuasive by also causing serious damage or a prolonged debilitating effect.

All this gets me thinking with regret about a forgone travel opportunity of my youth: "In Panama once, somebody tried to get me to pick up one of those big bullet ants, but I didn't. What did I miss?"

"You missed a fascinating experience," Schmidt says. "I guarantee you would not have forgotten it." These large ants get their name because the sting "feels like a bullet went into you," he says. His own encounter occurred in Brazil, when he was digging up a nest. "I knew right away. I said, 'WOOH!'" When he sat down to savor the experience, the hand where he'd been stung started shaking uncontrollably. He took a sharp pencil and, in the interest of science, used it to "poke pretty hard" in the afflicted area. It was numb.

"So that was interesting," Schmidt says, "but beyond that it was just unmitigated, excruciating pain." After a few minutes, he went back to his digging and got stung three more times. He was "still quivering and screaming from these peristaltic waves of pain" 12 hours later, despite the effects of beer and ice compresses. "This is the kind of science where you can never prove that one insect is the number one stinging thing," he says, "but at a certain point, you run out of viable competition.

"They're a big ant, they've got a lot of venom, and it's quite toxic. A small thing like a bird or lizard could get some serious physiological damage if one of these things really nails them." A small monkey might not die from the sting but if its hand starts to tremble, "that's not good for moving through the trees or avoiding predators. And that's how natural selection works."

Teaching people to stay away from a honeybee nest is an expensive process, if you have to do it one person at a time, and especially if each sting means the death of a bee. "But if you have a social organism like us," says Schmidt, "and if the family group sees Jill go into anaphylactic shock after a sting, everybody instantly gets the idea. You don't want to mess with this thing."

It may not be terribly comforting to realize, next time you suffer a painful sting, that you are a show-and-tell exhibit in some insect's continuing education program. But it could be worse. In one horrific case in Zimbabwe, a man disturbed a honeybee hive and was attacked so relentlessly that he had to jump into a river and hide beneath the surface. The bees continued to sting him every time he came up to breathe. They were so dense he had to suck bees into his mouth and chew them to get any air. The attack went on for four hours, producing diarrhea, among other systemic effects, so that he was passing bees out one end while still ingesting them at the other. Finally, nightfall drew the bees back to their hive and the victim dragged himself ashore. His face was literally black with embedded stings, and his hair was matted with dead bees. The doctors who treated him over the next few days counted 2,243 stings. But the victim lived to deliver the insects' familiar lesson plan, which was, as always: Leave the hive alone.

Freaky Friday

"And if a person does get stung, what do you advise?" I ask Schmidt, as I am walking out the door. A sissy question.

"There are various treatments," he says, "and basically all

they do is take your mind off the pain." He offers a few home remedies—spitting on the sting, or applying a wet pack of salt. But his heart isn't in it. For the serious student of nature, the point is clearly to savor the pain and pay attention to the sequence of physiological events—the depolarizing of pain receptors, the capillaries being constricted, the ruptured blood cells leaking hemoglobin.

"And if it's really bad, there's no harm in screaming?"

"Nah, go ahead and scream," he says. "Have a good time."

Just then, a tarantula hawk whirs through his yard at eye level. It's black, with metallic blue flanks, and about as menacing as a Chinook helicopter. "I can catch it if you want to get stung?" Schmidt offers. The thought of three minutes of totally unbearable pain is of course tempting. But before I can say, "Sure, it's Friday, let's go for a four," the wasp is out of reach.

Anyway, a man can only stand to have so much fun.

Life List

On a weekend not long ago, Russ Mittermeier, the peripatetic president of Conservation International, was visiting the dusty town of Farafangana, on the southeast coast of Madagascar. His official goal was to build ecotourism to what is variously known as "the eighth continent" and "the twelfth poorest nation on Earth."

Mainly, though, Mittermeier was in Farafangana to bag two new lemurs and add to his already vast primate life list. Science currently recognizes about 650 species or subspecies of apes, monkeys, and prosimians, 6 of them first described by Mittermeier himself. He has seen more than half of them in the wild, possibly more than anyone, ever. Among other things, he has sat nearby, wondering whether to avert his eyes, while mountain gorillas had sex. He has also stood watching nervously while chimpanzees ripped a live colobus monkey to shreds. Once, at a *bai*, or forest clearing, in the Congo, Mittermeier was watching lowland gorillas, when the gorillas sidled off, circled around, and sat down 30 feet behind, hidden by foliage, to watch him. (Maybe it was the start of a different sort of life list: Russ Mittermeier, *Homo sapiens.* Check.)

This trip to a couple of remnant scraps of forest 30 kilometers outside of Farafangana was aimed at adding to his list one new species, the white-collared brown lemur, and one new subspecies, the southernmost variety of the black-and-white ruffed lemur.

Mittermeier was dressed for the hunt in sneakers, white socks, shorts ("I like to see the leeches," he said), a khaki shirt, web vest, and a baseball cap, turned backward for better visibility while thrashing through the underbrush. Leitz 10 x 40 binoculars hung from his neck, beside a Nikon digital camera with a 400-millimeter vibration-reducing telephoto lens. At his belt was a Brazilian machete in a duct-taped sheath, and a Nalgene water bottle containing the murky leavings of miscellaneous Cokes and Oranginas, diluted with water. (On a tour when the animated film *Madagascar* was in production, Dreamworks executive Jeffrey Katzenberg professed horror at the "disgusting water bottle." But Mittermeier was unabashed: "I don't like to waste things.")

Up to now, life-listing has been largely an ornithological affliction, with birders trekking to the far ends of the earth to add some obscure feathered thing to their list. Mittermeier's college-aged son John, for instance, has already checked off 4,000 bird species, roughly double his father's bird list. "There are millions of Web sites for birders, and it's a multibillion-dollar industry," said Mittermeier. "So why not primates?"

He is promoting primate life-listing at least in part for the sheer joy of counting coup. Status competition is an important behavioral phenomenon in most primate groups, notably including the Mittermeier family. Among other things, everybody in the family keeps track of countries and "countrylike entities" visited;

they exchange cryptic e-mails: "37" or "49." Mittermeier said there is not much his eldest son can't beat him at these days ("I used to have chin-ups on him") but "he doesn't have my country list," which now stands at 114. There is a corresponding, though less celebrated, list of exotic diseases: "I've had leishmaniasis, schistosomiasis, pin worms, hook worms, everything else. No malaria. Lucky." But apart from hardship and the thrill of one-upmanship, what's the appeal of primate life-listing for other travelers?

Lemurs in particular swing through the treetops and pirouette along the ground more gracefully than any human ballet dancer. Primates also catch and hold the attention with behaviors that are a tantalizing mix of the deeply familiar and the foreign. And they are colorful, more so in some cases than birds. (The British naturalist Gerald Durrell once encountered a mandrill in full sexual display, its bottom like "a newly painted and violently patriotic lavatory seat," all blue on the outer rim and "virulent sunset scarlet" within. "Wonderful animal, ma'am," Durrell said to his guest, Her Royal Highness Princess Anne. Then he added, "Wouldn't you like to have a behind like that?")

But Mittermeier is promoting primate life-listing mainly with the idea that it will be good for the primates, by bringing ecotourism dollars to local people who might otherwise value forests only for fuel and building material, and lemurs only for meat. Tourism has boomed in the aftermath of the first *Madagascar* movie, to about 300,000 visitors a year. Mittermeier thinks life-listing could sustain that trend.

Lemurs are adorable, unlike, say, howler monkeys. "The best are *Propithecus candidus*," said Mittermeier of the species also known as silky sifakas. "They're big and fluffy with a pink face, and you think, 'This isn't a real animal. It's a Disney creation.'" (On the other hand, the aye-aye, *Daubentonia madagascariensis*, looks like something out of a Stephen King novel.) Though it's

standard to say Madagascar is as big as France, or Texas, deforestation means they actually survive in an area Mittermeier
described as more like "three New Jerseys." In some protected
areas, it's possible, with a little hiking, to see 10 different species. Visit five or six sites, and you can get 25 species in a trip.

So is the notion of primate life-listing practical? "There's an
obvious reason to be skeptical about the idea," says John Mitani,
a behavioral ecologist at the University of Michigan, who studies chimpanzees in Uganda. "Unlike bird-listing, it's not something you can do in your backyard." The United States has about
900 wild bird species anybody can get started on. Birdwatchers
can also buy field guides, birdfeeders, birdsong iPods, and even
video "nest cams" to sustain their passion year-round. There's
no equivalent for watching primates (unless you count eaves-
dropping on your human neighbors). Baboons and gray-cheeked
mangabeys don't, as a rule, turn up in our backyards. So the only
way for beginners to get started is to travel. "Here you're talking
about something that appeals only to people who have money,"
says Mitani. Then it dawns on him: "It could be good to have people with money interested in this."

But primatologists also express a philosophical reservation.
Though they tend to put it more diplomatically, the truth is that
they hate "twitchers," the sort of birdwatchers who check off a
species on a list, and then hurry away to bag the next species. A
name is just a label, says Amy Vedder, of the Wildlife Conservation Society. "I like actually watching animal behavior, the quality of what's going on in their lives. Being able to observe over
time is really where the magic is for me."

On the other hand, Vedder concedes that it was the twitchers who first got people thinking in the 1980s about how to save
a forest called Nyungwe in southwestern Rwanda. "There was no
visitors reception, no tour operators, and yet people were com-

ing. It was an indication of international interest, and of the special quality of that forest." In 2004, Nyungwe became a national park, protecting 600 square miles of spectacularly mountainous terrain. Vedder ranks it as one of the best spots in the world for watching primates. You can sit along a trail, she says, and "see large groups of colobus monkeys moving overhead for hours, because there are so many of them."

And you can do it more or less alone, a special attraction of primate life-listing. Birders tend to travel in packs and follow deeply beaten paths. Roughly 80 percent of people who watch *any* kind of wildlife are actually watching birds, according to the American Birding Association. But primate watching is still fresh territory. For instance, even biologists didn't realize until 1968 that chimpanzees, our closest primate relatives, live in stable groups and have complex social lives. The colorful sex life of bonobos, our other close primate cousin (and a species Mittermeier hasn't yet managed to add to his list), is also a relatively recent discovery. In addition to new behaviors, scientists discover new primate species at a rate of about one every year. With a third of all primates currently listed as critically threatened or endangered, Mittermeier's idea is that ecotourism could keep some of them from going extinct.

Have You Found the Animal?

At the Manombo Reserve outside Farafangana, the hunt took place in French, English, and Malagasy, with Mittermeier's host, biologist Jonah Ratsimbazafy, pausing now and then to cry out, *"Any ve ny biby?"* (Have you found the animal?) to various local guides. Ratsimbazafy also tried speaking to the lemurs in their own language, a back-of-the-throat, *gup-gup-gup* sound. Finally, one of the guides said, *"Aty!"* or "There it is!"

The white-collared brown lemur, *Eulemur fulvus albocollaris*, sat on a branch 30 feet above the trail, serene as a cat, with its long fluffy tail folded over one arm. It had big, tawny, mutton-chop cheeks, and its hazel eyes glowed in the afternoon sun. It gazed down curiously as Mittermeier crept underneath for one more close-up, and then another and another and another. A half hour later, the lemur was still there, the sort of trusting behavior that often leads biologists to regard them as not too bright. "They're not chimpanzees," said Mittermeier. "They're not even capuchins."

Next day, Mittermeier was back, accompanied by an official from the regional government, to look for the somewhat more skittish black-and-white ruffed lemur, *Varecia variegata variegata*. At a rickety log bridge, a big Mercedes truck was stuck up to its hubcaps, under a load of freshly cut trees from the reserve.

"It's illegal," said Ratsimbazafy. "But nobody cares out here in the remote forest."

"Except for today," said Mittermeier, as the regional official ordered the driver to report with his boss to the gendarmerie.

"But it's 99 days for the thief," said Ratsimbazafy, "and 1 day for the conservationist."

In the forest, the black-and-white ruffed lemur soon turned up walking on a branch overhead, looking like a small, arboreal panda. Mittermeier held up a tape recorder and played the call of another lemur, the indri, a high-pitched sound like air being slowly let out of a balloon. No reaction. Then Mittermeier played the chucking and squealing of other black-and-white ruffed lemurs. The animal looked up sharply, cocking its head one way and then the other to locate the sound.

"He's coming, he's coming to fight! Look!" said Mittermeier.

The lemur leapt into the tree directly overhead, considered the possibilities for a moment, and then moved off again. "He's chicken!" said Mittermeier, disappointed.

"He may have been feeling alone," said Ratsimbazafy.

On the walk back out of the forest, Mittermeier and the regional official talked about the potential of primate life-listing. "This is really a ticker thing to come here and get *Varecia* and *albocollaris*," Mittermeier said. "There are eight critically endangered primates in Madagascar, and we knocked off two of them here." *Albocollaris* in particular exists nowhere else on Earth. The regional official, an economist with a corporate background, worried that Farafangana was too far off the usual tourist circuit.

"Improve the road, that bridge mainly," said Mittermeier. "Clean up the trails, so you're following the route of the animals. You could do it in two weeks with a few guys. *C'est tres facile*." Next morning he made the same pitch to the owner of the local hotel, a former schoolmate of Ratsimbazafy's: "Each of these animals in the forest is worth a fortune to the region. If you eat it, you get a meal for one day. If you keep it in the forest, people will keep coming and coming to see it."

Later, flying out of Farafangana in a chartered Cessna 172, Mittermeier circled over the patches of forest he had just visited, dwindling remnants in the vast landscape of deforestation. But it was his nature to be optimistic: "You come to a place like this, meet the number two in the regional government and the big businessman, and soon people are saying, 'The *vazaha* came, the foreigners, and talked about how important the lemurs are,' and all of a sudden attitudes start to change."

Below, columns of smoke rose here and there where farmers continued to nibble the forest into oblivion. "This is basically it," said Mittermeier. "These two properties are the future for these animals."

Ghosts in the Grasslands

It was open season on babies in the Serengeti. Since mid-February, the vast antelope herds had been raining down fawns on the rolling African plains. They were tender, glossy-eyed creatures on unsteady legs. Their mothers bedded them down under grassy tussocks and whistling thorn acacias, hoping they would go unnoticed until they were strong enough to run with the herd. All a cheetah had to do was meander around like a child on an Easter egg hunt, nosing under the vegetation till it found a treat.

A seven-year-old named Talisker and her four cubs had just finished off a young Thomson's gazelle. Two of the cheetah cubs sat and rasped their long tongues up one another's cheeks, their heads preening with pleasure. When they walked, biologist Sarah Durant studied their bellies through her binoculars. The cubs ranked a 9 and Talisker an 8 on the Serengeti Cheetah Research Project's belly-fullness scale of 1 to 14, with 14 defined as "swallowed a basketball" and 8 being about hungry enough to start hunting again. In the distance, a column of sunlight broke through the slate-colored clouds. Sensing a behavioral pattern among field researchers, I asked if Talisker got her name from the whiskey. "A particularly fine one," said Durant. "Her sister's named Laphraoig." Talisker took up sentry duty atop a termite mound.

"There's a wildebeest calf over there," said Durant, as if hinting to Talisker. No one here roots for Bambi, except as Bambi tartare. Talisker soon spotted the calf, a tawny creature about the size of a horse foal, which had made the fatal error of becoming separated from its herd. The cheetah closed on the calf from the left rear flank at a trot, not bothering to stalk or slink.

When the wildebeest started to run, Talisker dug in behind. They turned, the cheetah's great doglike paws kicking up clots of dirt and grass. The distance between them narrowed. Still running, Talisker reached out and swatted the wildebeest across the haunches with her forepaw, bringing it down in a tumult of dust. The wildebeest struggled to its feet and reared up to shake off the demon now locked on its windpipe. Talisker went up on her hind legs and hung on. The four cubs came bounding up from behind and shoved the wildebeest back down again. They stood on its haunches while Talisker adjusted her bite for the slow strangulation. The wildebeest's hoofs flailed in the air. The cubs wandered off, looking vacantly away from the kill. "Waiting for mum to prepare the food," Durant said.

We waited too, and I reflected that I had come to a strange pass in my career as a natural-history writer. The predation was familiar enough, but I wasn't accustomed to such pretty killers. My inclination has always been to write about the less popular animals on Earth. The Serengeti is a place I routinely tell people not to visit, at least not until they've gotten to know the little killers in their own backyards. And yet here I was in the most glamorized habitat in Africa, writing about one of the most admired animals on Earth. It was a bit like a crime reporter being asked to profile a *Sports Illustrated* bathing suit model. I was appalled, and also ineluctably attracted.

A cheetah ambling across a field is among the most beautiful animals on Earth, long legged and slim, shoulders rolling, lithe as a fashion model on the runway. Then the cheetah breaks into its 60-mile-an-hour sprint, and it's as if Naomi Campbell has transformed instantly into Jackie Joyner-Kersee: The spine flexes up and down with each huge stride, the wide nostrils flare, the head weaves back and forth to fix the cheetah's big copper-colored eyes on its prey, and the tail stretches out to counterbalance the cheetah's weight on sharp turns. It becomes a blur of golden hair and dark spots, all speed and deadly finesse.

So why appalled? It had to do with human attitudes. Partly it was my own misguided notion that a creature as beautiful as the cheetah, the poster child of tire companies and tourist hotels, should need no help from me. And partly it was because cheetahs are cats, and they elicit the sort of cat-lover sentimentality that makes me cringe. One day a conservationist who ought to have known better pointed out the characteristic black lines running down from the inside corners of the cheetah's eyes and described the cheetah to me as "the cat that cries."

Something about cheetahs makes us patronize them. Maybe it's because they're the only big cats that do not attack humans, or because they're the only ones that cannot roar. (They growl and hiss. But they also chirp at one another like birds.) Humans have tamed cheetahs as pets and kept them as hunting animals for at least 3,700 years, and our ability to dominate them has given us the idea that cheetahs are only half-wild.

In fact, about 1,300 cheetahs now live in captivity and perhaps 12,000 in the wild. In my lifetime, they have vanished from Asia, except for a few in Iran and Pakistan, and dwindled to a rumor in North Africa. They are disappearing below the Sahara, too, as the growing human population encroaches on old cheetah habitat.

Parks provide some refuge, but most parks are overloaded with lions, which kill cheetah babies and steal cheetah kills.

Deadly Finesse

The Serengeti in northern Tanzania is one place to see how cheetahs live in the wild, untainted by sentiment. Over the past 25 years, biologists at the Serengeti Cheetah Research Project have come to know more than 400 individual cheetahs, using their distinctive spot patterns as identifiers. They've also constructed genealogies, stretching back in Talisker's case to her great-grandmother. Durant, who now heads the project, is a tall, handsome Englishwoman from the Zoological Society of London, strong-chinned and soft-spoken, with short blonde hair brushed up willy-nilly, and little round silver-rimmed sunglasses.

As we waited at the wildebeest carcass, Durant recalled meeting Talisker as a cub and later as "a hopeless adolescent." The cheetah's deadly finesse does not come easy, Durant said. Cubs spend about 18 months with their mothers learning how to survive, and it may take them another year or two to become good hunters. They often start out chasing wildly inappropriate prey, including buffalo. Talisker, said Durant, would hunt right in front of a hyena and then be surprised when it took away her kill. But she also learned from her mistakes. "Talisker's a pretty switched-on cheetah. She's quite vigilant and she moves away if she sees a lion or a hyena. She hides her kill where she can. She's a good hunter." She's also a successful mother, having already reared two litters to independence. Most females, said Durant, rear fewer than two *individual* cubs to independence in an average lifetime of seven years.

Talisker was now sitting off to one side, leaving her cubs to sort out the wildebeest. One of them touched it gingerly with a forepaw. Another stared down at it, then at his mother, then back at the carcass, like a kid confronting his first lobster dinner and hoping for help. "Wildebeest have thick skin, compared to a gazelle," Durant said. Cheetahs have small, delicate mouths. Their big eyes are set forward, the better to focus on prey, and the foreshortened face leaves little room for the mouth. One cub practiced a clumsy throat bite on the carcass. Another gnawed at the thin webbed flesh where the hind leg joins the belly.

The heat shimmered sideways across the open plain. The clouds became cottony and less ominous. When the cubs came up from the kill now, their mouths were blood-red. Talisker joined them, and the wildebeest's midsection seethed with golden fur and spots and black ears and sharp elbows. Sometimes two cheetahs fed opposite one another, with the tops of their heads pressed together like bookends. Sometimes they ate cheek-to-cheek and growled at one another over tidbits. After an hour, the cubs had clownish round bellies, and their faces looked like Kabuki warriors', covered with blood all the way back to the ears. They took turns eating now.

"She's about a 12," Durant said, eyeing Talisker's belly. Still room for topping off. "It might hurt, but she'll do it." Cheetahs normally get a kill only every one to three days, and they sometimes lose their dinner to lions and hyenas. So gorging makes sense. Three hours later, the cheetahs were still occasionally picking at their food, but mostly lying around, their bellies now spotted white hillocks rising from their midsections. Vultures had begun to edge closer, like waiters who want a banquet to be over so they can clean up and go home.

Loathsome Lions

One day Durant and I were talking about why cheetahs are so scarce, and I asked her, "Do you ever get to hate lions?"

"Surely everyone does?" she replied. "They are the thugs and bullies of the Serengeti. Of everywhere, really."

Cheetahs are not sturdy enough to defend themselves from either lions or hyenas, so they have become expert at being elsewhere employed. If their rivals hunt mainly by night, cheetahs hunt by day. If their rivals favor thick herds of wildebeest, cheetahs concentrate on gazelles, and hunt where the prey is less dense and there are fewer eyes to notice them slinking through the grass. The cheetah's famously swift chase lasts, on average, only 10 seconds, and brevity is a good thing. It means the flurry of a chase is less likely to attract attention. After bringing down her prey, a cheetah will typically lie still for several minutes, to recover her breath and also to check that no one is watching.

For a cheetah, the real danger is not losing a kill but losing her cubs. Ninety-five percent of cheetah cubs die before reaching independence. Hyenas kill them out of hunger, lions apparently out of bad habit. Durant theorizes that killing cheetah cubs is simply an extension of the male lion's urge to kill the cubs of any unknown lioness he meets, so that he can get her pregnant with his own cubs. Lionesses kill cheetah young too, to protect their territory. Female cheetahs deal with the threat by constantly moving, preferably before their rivals even know they're around. They coexist as a phantom species, slipping into temporary vacancies between prides of lions and packs of hyenas. Over the course of a year, a female in the Serengeti will typically wander an area of 320 square miles, larger than New York City. Several females may overlap in their wanderings, but, even so, cheetahs tend to be thin on the ground. Durant believes that there are no more than 250 cheetahs spread out across the entire Serengeti ecosystem, versus 2,800 lions and 9,000 spotted hyenas. The entire cheetah population of sub-Saharan Africa may have been small, even in the best of times.

If females wander the whole city, males stick to a few choice saloons on the Upper East Side. Because they have no cubs to worry about, they can maintain and defend small territories, averaging just 20 square miles. Two or three males, usually brothers, may hold a territory jointly. One day, Durant and I watched such a coalition, two males both pushing 12 on the belly-fullness scale, prowl through the bush, stopping now and then to gaze at antelopes and wildebeests. "I often think they watch prey the way we watch television," said Durant, "because it's comforting and mindless."

The cheetahs headed for the tallest tree in the neighborhood, an umbrella acacia, and squirted urine on the trunk at roughly cheetah nose height—a message that, according to one biologist, tells rival males, "If I catch you in my territory, I will kill you." It's also a way to attract the attention of passing females. Durant went over to inspect the acacia. "Oooh, bliss," she said, spotting a fresh scat nearby. She scooped a sample into a small laboratory bottle, to be sent off for DNA testing.

Paternity is one of the great unknowns of cheetah biology, not just for researchers but also for the cheetahs themselves. A female's home range may contain three or four male territories and she may mate with any of the resident males, as well as with floater males that pass through. Durant has seen cheetah mating just once. It involved a coalition of three males named Daniel, Day, and Lewis, after the actor, and a female named Florence. All of them disappeared into the bush. After a few seconds, Durant saw what she called "stacked cheetahs," Daniel, Day, and Lewis mounted one atop the other, with Florence "looking rather squashed" on the bottom. But confusion can be a good thing. Unlike lions, male cheetahs have never been known to kill cubs, perhaps because they have no way of knowing whether the cubs are their own offspring.

The question of who fathers the cubs is of special interest because cheetahs are a genetic mystery. In the 1980s, researchers discovered that all cheetahs are genetically similar—so much so that skin grafts from one cheetah to another produce no immune reaction. The finding caused geneticists to rethink the cheetah's evolutionary history: Roughly 20,000 years ago, cheetahs ranged around the world. At different times, there were two species in North America alone. But cheetah populations apparently suffered a drastic decline about 10,000 years ago, and all cheetahs now living appear to be descended from a relative handful of survivors. No one really knows what this signifies for the future of the species. Some biologists suggest that, having survived the population bottleneck and recovered, cheetahs in the wild suffer no ill effects from their genetic homogeneity. Others believe that they may be unusually vulnerable to any small change in their environment, particularly disease. Either way, the cheetah is a conundrum: The fastest animal on land, an apparent model of evolutionary fitness, is also as inbred as the average lab mouse.

"The more information you get, the more fascinating an animal becomes," Durant said one day. "It doesn't matter if it shows that an animal is more peaceful than we thought or more aggressive. It's the information itself. Anything that makes people value an animal for what it is, rather than for our fantasy of what is, the better it is for the animal." I had a small problem figuring out how to value the Lion King as a cub killer and a kleptoparasite. But cheetahs were beginning to grow on me.

The Utilization Process

I went to Namibia, on the southwestern coast of Africa, where people were advocating a different, distinctly utilitarian, view of cheetahs. The world's largest wild cheetah population lives here

Riding the 90

I was working on this story with wildlife photographer Chris Johns and Dave Hamman, a South African who was our guide, driver, and photo assistant. For part of the field time, we stayed at Mombo Camp on Chief's Island in Botswana's Okavango Delta. It's a classic safari lodge, which is to say we were hopelessly spoiled. I slept in a platform tent with a little front porch looking out onto the floodplain. There was a bathroom enclosed with reeds on the back deck, and the end of the toilet paper was neatly folded and held in place with an acacia thorn. The vervet monkeys didn't get it. "Wait! Look!" I told them. "This is what evolution can do for you." Instead, they came bouncing down the tent fly, stole the toilet paper, and shat in the sink.

The tent had three windows on each side, with white linen curtains tied back. There was a lantern-style lamp on the night table, a fan overhead, and an armoire for nervous tourists to hide in should a leopard drop by for a late-night snack. The first night, a baboon not far from camp gave out a braying, donkeylike alarm call in the middle of the night. Lions killed a zebra nearby around 2 a.m.

Next morning we headed out at first light, as the ghostly shapes of kudu on the floodplain were still beginning to take on substance. A couple of spotted hyenas passed us heading back toward camp, on their way to last night's kill. (Too late: The only thing left of the zebra was a patch of blood.)

A few miles down the road, Dave stopped the vehicle and peered out in the direction of a birdcall. "That's a francolin, that chickenlike bird," he said. "When it makes that alarm call, 80 percent of the time it's a leopard." But before we could turn to follow, a guide radioed in a sighting and we went lurching off in the opposite direction.

We were using a Land Rover Defender 90, an open vehicle

good for viewing wildlife and taking pictures. Chris sat in the passenger seat, where he had a camera arm mounted. I was in back, under a low ceiling, leaning against the spare tire on the rear gate, with my legs bent and my feet jammed against the roll bar supports. Then the gate came flying open behind me. So I switched to sitting sideways, holding it shut while trying not to bounce too often into the steel roll bar overhead.

I quickly gave up riding in the back of the 90, except when we were trying to get somewhere fast. Instead, I sat on top, above the front seats, with my feet propped against the windshield roll bar. There was a lot of side-to-side lurching up there, but the handholds on the roof helped, and the view was splendid. It only got awkward now and then when a hole appeared unexpectedly out on the floodplain and the 90 went thumping down to an abrupt halt. Or when we drove in close to a pride of lions and one of them looked up at me as if I might be an item on the menu.

Chris, Dave, and I got along brilliantly, with none of the usual writer-photographer, cat-dog bickering. Our hunting technique was to drive around in the 90 listening for francolin alarm calls, checking out clouds of dust, and studying promising shapes in the shadows of distant termite mounds. Sometimes, waiting around for a cheetah or a leopard to get up and do something, we just sat and talked—about the stupidity of institutions and editors, or about spending too much time away from our families, missing too many ball games and concerts back home. We had a continuing conversation about dealing with resentful wives, parts one, two, and three (then again, we were giving special meaning to the phrase "remote, incommunicative husband"). A pregnant cheetah, awkward and sore-bottomed, reminded Chris of his wife when she was about to give birth. A sausage pod tree where we waited for a leopard one day reminded Dave of local folklore: You take sap from the young seed pods and apply it to an incision in your penis. Then when the fruit grows long and round, your penis will, too. "I was talking to a doctor and he said, 'You know, that would work,' and I thought, Yes, we're onto something worth millions here. Then he said, 'With the infection and the swelling.'"

Back at the lodge that night, a couple of women tourists were having too many drinks and one of them finally said to us, "Now the fellow with the vacuous expression on his face is the photographer? And the handsome fellow with the beard who never talks to anybody, he's the writer?" Chris blanched. I beamed.

Well, OK, the cat-dog thing was never too far below the surface.

and thrives largely because landowners have exterminated lions from the huge private ranches that dominate the countryside. Unfortunately, the ranchers mostly regard cheetahs as vermin, too; they're just not so easy to get rid of. Namibia has about 3,000 cheetahs, down from 6,000 in the 1980s, and many ranchers argue that the best hope for saving the cheetahs is to let wealthy foreigners trophy-hunt a small percentage of them each year. It's a measure of how tangled and difficult the question of cheetah conservation has become that even some environmentalists say the ranchers are right.

Traveling Namibia's excellent two-lane blacktop highways, I frequently had the strange sensation that I'd just woken up in Wyoming. The highway rolled for hours through a flat landscape of parched grass and tangled gray scrub. Red rock reared up in knobbled sandstone ridges and strange, fanciful promontories, here a pyramid, there a planetarium. On the car radio a steel guitar twanged country and western: "I think I'm on a roll here in Little Rock." And then the music stopped, and a disk jockey from the Damara tribe delivered his patter in a language full of popping and clicking, a sound like soda cans being opened and chicken

being sucked from between the teeth. On the next station down
the dial (a dial of vast empty spaces) the announcer was sputter-
ing German and playing lieder songs.

The descendants of European colonizers, mostly Germans
and Afrikaners, make up only about 5 percent of the population
in Namibia, which won its independence from South Africa in
1990. But they still dominate the rural landscape. Their lovingly
tended desert towns have rows of palm trees planted in the medi-
ans and little churches lifted intact out of the Bavarian countryside. Their long fences divide the desert into vast, arid ranches
where antelope roam, along with cattle and cheetahs. Almost all
the ranchers I visited were dyspeptic on the subject of predators,
the way American ranchers talk about wolves or grizzlies.

At his 19,000-acre ranch south of the Waterburg Plateau,
Tinus van Rensburg came out to greet me and immediately held
up his left hand to ask if I'd brought him a new finger to replace
one bitten off a few months before by a cheetah. He was wearing
a surplus German Army overcoat, and he had scraggly dark hair,
a thick beard, a gap-toothed grin. Inside we sat by the fire with
an old cheetah-skin rug beneath our feet and a caracal hide covering the table. Van Rensburg began to recite the basic problem
with cheetahs: In short, they eat the same meat we do and often
get there first.

As a cattle rancher, van Rensburg said, he used to lose 15
calves a year to predators. The problem only got worse when he
abandoned cattle and converted his ranch to trophy hunting.
He stocked his ranch with black-faced impalas, springbok, and
blesbok, and the cheetahs ate almost all of them—worth $24,000
(U.S.) in the previous three years.

"Oh!" he said, "my heart pumps blood. One day you feel that
the cheetah also has a right to stay here. But you get angry, and the
moment you see a cheetah, you start shooting. I think you would

feel the same." I asked him how many cheetahs he typically took off his land. "Oh, that's a lot," he said. "Last year I think I catch and shoot 17."

The accident that cost van Rensburg an index finger happened when he released his grip on the neck of a cheetah he had trapped and it spun around, catching his finger in its side teeth and holding on. "I could do nothing. I must sit and wait till he is finished," van Rensburg said. He killed the cheetah. But he meant to suggest that his antipathy toward the species was merely practical, not pathological, not Ahab after his white whale. Of the finger, he said, "I think that was my own stupid thing to do. But I blame the cheetah for the small game, because I've lost a lot of money."

The ranchers I talked with readily volunteered that their cheetah problems were at least partly of their own making. "There's far more cheetahs today than there were a hundred years ago," one rancher said. Back then, when the water vanished during the dry season, the grazing herds went elsewhere, along with their predators. But the ranchers put cattle watering holes everywhere, inadvertently making it possible for game and cheetahs alike to remain in the desert year-round. The ranchers also wiped out hyenas and lions, leaving cheetahs and leopards as the top predators. In the 1980s, Namibia's cheetahs were thriving on a growing population of antelope, particularly kudu.

Then the kudu went bust because of drought and disease, and the cheetahs increasingly turned to livestock. Ranchers were allowed to kill cheetahs without limit, and the government estimates that they took as many as 600 a year. "They were required to report kills," a professional hunter named Volker Grellman recalled. "But most didn't. They threw 'em into an aardvark hole and covered 'em up." The killing has since slowed to about 180 cheetahs a year, according to a Namibian wildlife official, though he added that the number may still be "way underreported."

None of this sounded to me like an argument for trophy hunting. But that was one of the main strategies being discussed for cheetah conservation, as a way to make cheetahs valuable to the ranchers who have to live with them. The Namibian government and several hunters had applied to the U.S. Department of the Interior for permission to import up to 50 trophy cheetahs a year. U.S. permission matters because American hunters are willing to pay big trophy fees for cheetahs. (The petition was later denied.)

"The fate of the cheetahs is in the hands of the ranchers," said Grellman, a Hemingway look-alike with a broad face, a white beard, and blue eyes peering over black reading glasses. "You have to appeal to their goodwill and offer them something in return to compensate them for their losses." If ranchers could invite hunters onto their property, he said, and charge them a $1,000 trophy fee to kill a cheetah, with 15 percent going to a conservation fund and the rest split between the landowner and the hunting guide (often the same person), then cheetahs might become an asset instead of a nuisance. The parsimonious mind-set of ranchers strongly militates against flinging $1,000 animals into aardvark holes. Trophy hunting has worked with other species. The grazing herds of wildlife are once again abundant, Grellman said, because ranchers protect them as a cash crop. Cattle sell for about $300. But a kudu is worth $900 just for its head, and as one incredulous old rancher put it, "You get to keep the meat." The cheetah, he said, must "become part of the utilization process."

One evening I went out to watch a tame cheetah getting his exercise in a huge hayfield, with the long red bluff of the Waterberg Plateau lit up by the setting sun in the background. The exercise lure, a bright red rag on a ground-level wire-and-pulley system, took off, and the cheetah lit out in pursuit till he pinned it down with his forepaws. Then he lay panting helplessly on the ground. He allowed me to handle the tire-tread pads on his feet and the cleatlike, semiretractile claws. I felt the curved dewclaw on his foreleg, with which cheetahs snag and trip their prey when they swat out at the end of the chase.

The cheetah's owner, American conservationist Laurie Marker, was explaining to a group of Namibian agriculture students that cheetahs are relatively weak killers. They can't snap a victim's neck with their teeth, as leopards do. The adaptations that help make them such fast, efficient hunters also make them vulnerable at the moment of success to more powerful cats, like the lion. The students regarded the animal warily. Marker treated the cheetah like a pet. But to them it was still a predator, the enemy.

Back in the 1980s when Laurie Marker first heard how many cheetahs were being killed each year in Namibia, her impulse was to save the animals by catching them and putting them in parks or zoos. She'd made a career as a specialist in captive breeding of cheetahs in the United States. But Marker began to think that zoos could be part of the problem, as well as the solution. So in 1991 she moved to Namibia and formed the Cheetah Conservation Fund (CCF) with the aim of saving cheetahs in the wild.

Marker, who has dark ringleted hair framing big tinted eyeglasses and thin lips, had little experience with wild animals or ranchers. She is a lover of dogs, goats, and house cats, all of which she addresses affectionately: "Aren't you the most beautiful girl? Yes, you are." She talks almost the same way to visitors to her cheetah education center on a farm outside the town of Otjiwarango. Moreover, she does not speak German or Afrikaans, the languages of the ranchers. Not surprisingly, the ranchers gave her the sort of warm welcome customarily reserved for Americans arriving in foreign countries to tell the locals how to behave. "In the beginning," Marker said, "I thought we were going to get shot."

But her arrival in Namibia, and the cheetah survey she began to conduct among the ranchers there, sent a message: The outside world, which otherwise scarcely knew the place existed, actually cared about Namibia's cheetahs. At their ranch on the other side of Otjiwarango, Wayne and Lise Hanssen soon formed a homegrown cheetah and leopard conservation group called AfriCat. Though the egos on both sides frequently clash, AfriCat and CCF take the same approach of gentle persuasion with hostile ranchers. "We never say 'Don't kill cats,'" Marker said, "because the door just closes."

Wayne Hanssen, a classic rancher with a red mustache, a bush hat, and a skinning knife at his belt, won his first reluctant convert in his father, who had been losing more than 20 calves a year to predators and shooting any cat he came across. "He said he'd stop shooting when we got the loss down to five calves a year," Hanssen recalled, sitting on the hood of his Land Rover. To reduce predator losses, the Hanssens tried using guard animals, building better fences, and penning up the calves at night. After 10 years, his father could no longer "explain why he shot those cats. We proved by intensive farming methods that we could minimize the losses."

Most ranchers, Hanssen quickly added, aren't willing to make the effort. But the campaign of gentle persuasion appears to have produced at least one significant change: Having trapped a cheetah, many ranchers are no longer so eager to kill it. Instead, they call AfriCat or CCF to collect the animal. AfriCat was originally paying up to $250 for cheetahs, but both groups now oppose payment.

Catching Cats

One day I went with Lise Hanssen to retrieve a trapped cheetah. Three generations of the von Oppen family and their farmworkers drove out to the trap in a festive caravan. Jennifer Lee von Oppen, a nine-year-old, was planning to do a school report about cheetahs, and the last thing anybody wanted was to see the cheetah dead or to break up a cheetah family. Hanssen had picked up one cheetah from the von Oppens a few days earlier, and now two more cheetahs waited outside the trap. They fled as we approached, leaving their trapped brother to pace nervously in the cage. He spat when Hanssen came near, and Hanssen spoke to him soothingly, "I've got your mama at home."

Hanssen aimed a long, aluminum blowpipe at the caged animal and quickly fired a tranquilizer into his flanks. Then the children carried the unconscious 70-pound cub to a truck, for a medical examination and the trip back to Hanssen's ranch, where he would rejoin his mother. By the end of the week, the whole cheetah family was reunited in captivity, and after a six-week quarantine, Hanssen released them into a game reserve as part of a large-carnivore reintroduction program.

One problem with this warm scenario is that trapping can become a feel-good way to eliminate cheetahs in the wild. In South Africa most of the cheetahs have already vanished from private lands into zoos and game parks, which can be a reproductive dead end because of the cheetah's complex mating behavior and the potential for inadvertent inbreeding. Most U.S. and European zoos now swap animals and cooperate in other ways to maximize reproduction by captive cheetahs. But South African zoos and game parks have been slow to get involved, according to Laurie Marker, who manages the International Cheetah Studbook. Many of them treat captive-born cheetah cubs, which sell for $6,000 apiece, as a lucrative business. "When they have animals that don't reproduce, they call up farmers in Namibia and say, 'Please, will you catch some cats?'" Marker said. "Most want females, and for every female caught, the farmer typically catches 10 or 15 males."

"The perception is that zoos are saving the cheetahs because they're taking problem animals that would be shot," said Bonnie

Schumann, a South African on the CCF staff. "But they aren't just taking out problem animals. They're creating a problem by paying farmers to open their cages and catch cheetahs."

Despite the apparent success of captive breeding programs at some zoos, no one is now attempting to reintroduce captive cheetahs to the wild. There are relatively few wild places left in Africa or Asia big enough to accommodate cheetahs or willing to accept them. Thus both CCF and AfriCat emphasize protecting the existing population in Namibia and persuading ranchers to tolerate cheetahs in the wild. Despite their cat-lover orientation, both groups have come to support trophy hunting as one practical way to accomplish this. Hunting, said Marker, "won't make anybody rich, but it's an aspect of management."

I went to visit a young hunting guide named Jochen Hein at his game ranch near Okahandja, where a pet cheetah named Maggie patrolled outside the kitchen door. As we toured his ranch, Hein told me about a trophy cheetah one of his clients had recently shot on a neighboring farm. The neighbor, he said, used to kill cheetahs as pests. Then Hein paid him the cheetah trophy fee, and the farmer's jubilant wife confessed that it was their first income of any kind in six months.

"When we killed the cheetah," Hein recalled, "I said, 'Look, you're getting this money, keep those cheetahs for me. Please forget these damned traps.' A lot of these farmers, they need to see it for themselves. They need a professional hunter who will shoot a cheetah on their property and give them some money."

Hein helped write Namibia's new cheetah hunting compact, which sets the terms for an ethical hunt. But Sarah Durant doubted that the compact would protect cheetahs. It's too easy, she said, for cheetahs to wander from ranches that observe the compact to neighboring ranches that continue to trap and shoot.

Some hunters also objected to the new rules. Volker Grellman figured that a visiting hunter who spent 14 days in the field following the new rules would have only a 20 percent chance of even seeing a cheetah, much less getting off a shot. No landowner was likely to get more than one or two cheetah trophy fees a year, and the fees would never equal what the rancher lost from having the cheetah there in the first place. I repeated these criticisms to Hein, who replied that hunters would spend those 14 days taking kudu and other common game (and paying trophy fees on them), drawn on by the elusive prospect of a cheetah.

I said goodbye and headed from Hein's ranch to the nearest gas station, at an empty corner in the middle of the desert, where I happened to find the very farmer to whom Hein had recently paid the cheetah trophy fee. He was a silver-haired, dark-eyed man of about 50 in a flannel shirt and a leather jacket. When I asked him if cattle ranching paid a decent living these days, he laughed softly. He told me the usual story about losing 20 calves a year to cheetahs. I asked him how he felt about them now that he'd earned his first cheetah trophy fee.

"When I see them," he said, "I shoot them."

Maybe it was just bluster, a rancher saying what his father and grandfather always said. Or maybe old animosities do not vanish in one transaction, or even in one generation. "I can't honestly say we've ever turned around a farmer so that he never shoots a cheetah again," Wayne Hanssen had told me. Then he added, 'We've got a lot of farmers thinking twice."

The Wait-a-Bit Bush

Near the end of my visit in Namibia I went out one morning in search of radio-collared cheetahs with a French expatriate bush pilot named Jack Imbert. We took off from the dirt road outside the CCF farm, with radio-tracking antennas clamped onto both wing struts of his Cessna 206. Imbert's technique for pinpointing an animal's location was to fly figure eights in the vicinity of the radio-collar signal, banking so steep and so low that the entire port window was filled with the thorny earth skidding past just beyond the wingtip, and the starboard window was all empty blue sky. Then the plane seesawed over, and we did it on the other side. I got the feeling that the hairy love grass and *wag-'n-bietjie*, or "wait-a-bit" bush, was going to reach out and haul the plane into the tangled undergrowth.

We leveled off just above the ground and cruised past two cheetahs seated alongside a fence. One of them looked at this winged apparition, and his eyes burned with a color like the embers of a banked fire. Then he turned away with magisterial feline indifference. In truth, it was almost contempt.

I contemplated the trade-off: Those eyes replaced with glassy baubles. That head mounted on some trophy room wall, to be admired by cigar-waving partygoers and to become cloyed in time with cobwebs. The idea that one of the most beautiful animals on Earth needed to become "part of the utilization process" stuck in my heart. And yet 50 or so such cheetahs turned into trophies each year might just make one or two ranchers think twice. And maybe this was a beginning.

The plane flew on across the desert, and it seemed to me that nature seldom offers easy or reasonable trade-offs. She is content merely to teach us one hard lesson, over and over, and nowhere more vividly than in Africa: All life comes from death. Below us wildebeests angrily tossed their manes, and springbok blithely grazed. Somewhere in the thorny brush a cheetah ambled, doubtless thinking, as we all must, about where it would find its next meal.

The Enemy Within

At dusk on Jackson Square, in the heart of the French Quarter, Lola sat at her folding table waiting to read palms. A tall bearded man in leather and silver jewelry jangled past, and a pale woman in a Clara Bow haircut ghosted through the shadows. The chimney swifts swooped like bats above the steeples and lace balconies of the square, chittering with anticipation: tonight, or maybe tomorrow, a spectacular mass mating ritual would take place on the streets of New Orleans.

But there would be no joy in the Big Easy. Sybarites would react instead with horror and dismay. Foot traffic would stop. Restaurants would close their patios and slam their doors. Baseball games would be canceled. Bourbon Street would go dark. The scandalous event was the Mardi Gras of the Formosan termites, a mating flight in which insects by the gazillion fill the moist air seeking love. On certain evenings in May and June, the termites come seething out of the woodwork and trees in such tropical profusion that it can be difficult for a person to breathe, or for a driver to see halfway across the street. The alates, as the winged termites are called, are frail creatures, a third of an inch

long, fluttering weakly on their dull, diaphanous wings. They
have only a few hours to find a mate and set up house together.
More than 99 percent of them will fail, getting eaten by the
chimney swifts, or piling up like dead leaves in the corners. And
yet they induce fear.

The termites swarm toward light, even a television set kept on
by people huddled in their darkened apartments with the doors
and windows shut. "Any house with cracks, if you have the lights
on, they start coming in," said a resident of the French Quarter,
"and all the houses down here have cracks." The alates get caught
in peoples' hair and fall into their food. They crawl under collars
and creep across bedsheets.

But the real fear was that a few of the termites out flying on
this night would succeed: they'd find mates and settle down to
reproduce behind the bathroom wall or any other moist interior
surface. They'd set up housekeeping in some perfectly restored
Italianate town house on Royal Street, or in a double shotgun cottage on St. Peter Street, where they could do enough damage to
bankrupt the unsuspecting owner or drive a family to divorce.
Formosan termites, introduced into this country only about 50
years ago, are one of the most destructive pests in the world.
"There is this psychological terror they impart," said a French
Quarter resident, as he battened down in anticipation of swarming night, "because we know this is part of the life cycle of a creature that is eating our homes."

When I first heard about the Termite That Ate New Orleans,
I happened to be in the middle of building my own termite food
court—a new wood-frame house on the coast of Connecticut—and
it seemed like a good time to find out more about the enemy. I
follow entomological news the way other people read the sports
pages or the funnies, and what I knew about termites was a grab
bag of odd facts clipped out of newspapers over the years. I knew,

for instance, that there are about a thousand pounds of termites for every man, woman, and child on Earth, and that the termite biomass in some parts of Australia outweighs the kangaroos. I knew that up to 3o percent of the gases implicated in global warming may actually come from termite flatulence (I just wasn't quite sure why). I also knew that different termite species have evolved soldiers of wonderfully bizarre form: for instance, one type of soldier has a nozzlelike appendage on its head that sprays sticky gunk on ants, which are a common enemy of all termites. Other types have mandibles that work like baseball bats, popping ants out into short center field. I knew that, despite their reputation for destruction, termites are among the most formidable builders in the animal kingdom—and that what we call destruction is actually, seen from a larger perspective, recycling. If termites weren't out there breaking down dead trees and other plant material, and returning nutrients to the soil, new trees could not live, nor could we.

"As homeowners we tend to think of termites as pests," writes David Nickle, a U.S. Department of Agriculture entomologist, "but termites really are one of the most beneficial groups of insects on Earth." So I decided to learn more about termites, mapping out an itinerary that included stops in New Orleans and in southern Africa, where I'd previously seen mounds twice my height built by termites weighing less than a thousandth of an ounce each.

Disturbing Undulations

With termites, keeping a hold on the larger perspective can of course be a struggle—for instance, when you are stuck with 38 children in a classroom full of swarming alates at the Dr. Charles R. Drew Elementary School in New Orleans. In Monica Wilson-Leary's fifth-grade class, the termites had eaten wormy holes through the wood-backed blackboard, and she no longer posted her students' work on the corkboard overhead because termites would critique it one bite at a time. A month ago, she'd put up a strip of orange paper as a decorative border, but it was already buckled and swollen from the steady tunneling underneath. Sometimes it undulated faintly. In the corners of the room, the hardwood floors were shredded to papery strips. Wilson-Leary had gotten in the habit of sweeping dead termites off her desk every morning in swarming season.

Outside, Gregg Henderson of the Louisiana State University Agricultural Center was drilling holes in the dirt around the perimeter of the building and burying monitors—rolls of corrugated cardboard in containers the size of a soda can—to measure how bad the termite problem was. Later, exterminators would set out baits laced with newfangled poisons, including one that prevents the termites from molting and another that inhibits their cells from using energy. Then Henderson's crew would come back to see how many termites survived and how quickly they came back. The aim was to find the best poison for controlling Formosan termites in the absence of chlordane, a potent termite-killer that was banned in 1988 because it accumulates in the flesh of higher organisms, including humans.

It was a daunting challenge. Henderson said that since 1989, when researchers first started monitoring the mating flights in New Orleans, the annual catch of alates had increased 3,600 percent. Sometimes a single trap the size of an ordinary household bucket would catch 15,000 alates in a night. And the future looked even bigger and brighter. According to Henderson, if a male and a female found each other on their mating flight, it might take them seven years to build their colony large enough to produce its own alates. But after that, the colony could grow and send out reproducers at an exponentially increasing rate. A drought had caused a major decline in the city's termites in 2001; so did Hurricane Katrina in 2005 ("drowned 'em," said Henderson). But they seemed to bounce right back.

Thus it had occurred to some people that Formosan termites might be on the verge of a breakout. They have been building a beachhead in this country since they first arrived as stowaways on military surplus returning from the Pacific after World War II. Their territory now extends from Louisiana to the Carolinas, and also includes Hawaii and California. Some researchers have speculated that with the help of central heating, they could spread as far north as Boston, Chicago, and Oregon.

Beneath the Shiny Surface

They have a bleak saying in New Orleans, but it's true for much of the rest of the world, too, especially in warmer climates: There are really only two kinds of houses—the ones that *have* termites, and the ones that *will* have termites. In the United States alone, Formosan termites cause an estimated $1 billion a year in damage. (The estimate dates back to 2000 and does not include other termite species. So the actual cost may well be much higher. Henderson has found evidence, for instance, that termite damage may have been a factor in the breaching of levees and flooding of large portions of New Orleans during Hurricane Katrina.)

Termites generally have an aversion to open air, so, except for their annual mating flight, the signs of invasion are usually subtle. Subterranean termites build mud tubes up the house foundation, to get back and forth between their food and the nest. But unless a homeowner happens to be awfully zealous with the weed-whacker, these tubes are easy to overlook. Drywood termites actually nest inside the wood itself, sealing themselves off from the outside world to do their dirty work; they even manage to "drink" water out of old wood. They intermittently open "kick-out holes" to heave out their fecal pellets. But few homeowners recognize the source of the telltale grit scattered on nearby floors—even when they slip on it and land flat on their backs, staring up at the woodwork in which the termites have once again sealed themselves snugly away. (Some drywood termite soldiers have heads like drain plugs, to block up the kickout holes.) Another sign of invasion, also easily missed, is the faint clicking sound produced by some termite species. Europeans used to believe this sound was a harbinger of death in the house, which one expert thinks is how termites got their name, from the Greek word *terma*, or "the end." But more often than not, termites go unheard and unsuspected.

To protect themselves from the open air, they leave the surface of whatever they eat intact—until the chair legs collapse or the joists buckle or, as happened to a New Orleans homeowner, all four legs of the grand piano plunge through the floor. Moreover, when the surface finally crumbles to reveal the corruption underneath, termites are also ugly, with their maggot-pale abdomens dragging listlessly behind them. The normal human impulse is to reach for the nearest deadly aerosol and kill everything in sight. But even killing a million termites may not solve the problem; with enough time and food, that colony behind the pantry wall could number 5 million individuals.

Early writers sometimes mistakenly described termites as "white ants," perhaps to make them seem more appealing. In fact, termites evolved more than 140 million years ago from an ancestral species that also gave rise to the cockroaches, and they still share many physical traits with cockroaches. But while they are only distantly related to ants, bees, and other hymenopterans, termites somehow also developed roughly the same highly cooperative social structure. As in most ant species, a few indi-

viduals do the reproducing and the rest serve the colony by tend-
ing juveniles, gathering food, building the nest, or battling off
intruders.

For ant workers, giving up their own reproductive potential
for the good of the colony seems to make evolutionary sense, at
least according to kin selection theory. Because of a quirk in the
way ants reproduce, worker ants are typically closer to their siblings than they would be to their own offspring. But it's different
for termites; genetically, their offspring are as good an invest-
ment as their siblings. And yet most termites also give up the
chance to reproduce, and termite soldiers routinely defend the
colony with their lives. In one spectacular display of self-sacrifice,
the soldiers of a termite species swell up their abdomens till they
explode, spattering their guts all over any adversary that threatens the colony. To scientists, this constitutes highly social behavior, and there are many theories about how termites got it.

In Barbara Thorne's laboratory at the University of Maryland,
each of 500 clear plastic containers the size of a cake pan holds
an entire colony of dampwood termites, living out their lives in
a chunk of wood. "They are among the most primitive living termites," Thorne told me. "They spend their entire life cycle in
the wood. They don't go off foraging, like many other termites."
Thorne theorizes that early termites like these lived slow, isolated lives snug in their food source. The breeding pair, or king
and queen, were monogamous, like modern termites, and did
all the reproduction in the colony. Their offspring were capable of developing into alates, flying away, and becoming kings
or queens in their own colonies. But because cellulose is low in
nutrients, it took them months to reach maturity. This meant
that older siblings were still in the nest, available to help feed
and groom their juniors. Helping with brood care was the crucial
first step in becoming a family and then a society.

Dinner with the Family

Food sharing is of course commonplace in animal societies. For-aging ants, for example, bring back food and regurgitate for their nest-mates, who regurgitate in turn for other nest-mates, until everyone in the colony is fed. But termites take this communal feast a startling step further: they share food from both ends of the digestive tract. One can just imagine an appalled termite par-ent catching the youngsters at this unsanitary practice and declar-ing that they had better have a damned good explanation. And in fact they do: the "hindgut fluids," as scientists delicately call them, contain intestinal protozoa and bacteria that have evolved with the termites to help them digest dead wood. Plant cellulose is extremely high in carbon and low in the nitrogen animals need to build cells. The protozoa and bacteria help burn off the excess carbon (incidentally producing greenhouse gases) and concen-trate the nitrogen. Passing these fellow travelers on to each new generation is one of the vital functions of termite society. Some scientists suggest it is the reason termites developed a society in the first place.

In the course of evolution, the hindgut fluids also became one way for the termites to pass on pheromones, or chemical mes-sages, vital to regulating the life of the colony. One of these chem-ical messages, called juvenile hormone, is produced by the queen to suppress sexual development in some of her offspring and keep them around the house as helpers, usually for their entire life span of perhaps one or two years. Other chemical messages cause some workers to transform themselves permanently into soldiers, their heads becoming big and thick to support heavy weaponry. The workers and soldiers also contribute their own pheromones to regulate their ranks. For instance, foragers who find food lay down a trail pheromone so their nest-mates can find it, too. Every individual is part of a sophisticated biochemical feedback loop for meeting the needs of the colony.

Scientists don't yet understand what triggers the colony to produce its annual flight of alates. As the process begins, a large number of soldiers appear in the nest. Henderson theorizes that some of the soldiers sponge up the queen's juvenile hormone, allowing one crop of her offspring to escape the usual chemical sterilization and become alates. Guarded by soldiers and sometimes actually shoved out of the nest by their nonbreeding siblings, these alates fly away to propagate the colony—or die.

Having evolved social castes to handle different chores around the nest, and the intestinal microbes to exploit one of the most abundant but indigestible materials on Earth, dead wood, termites made one other evolutionary leap: instead of living in small groups within the narrow confines of their food source, they began to form more populous colonies and build elaborate nests, becoming the most prodigious engineers in the animal kingdom. The nests of some termite species contain perfect arches, spiral staircases between levels, even vent holes with gargoylelike structures. Some species nesting in rain-forest trees build "umbrellas" overhead. Other species build mounds with a north-south orientation, apparently for better climate control. Such adaptations have enabled the 2,300 known termite species to nest on every continent except Antarctica, and to function not just as wood-feeders, but as eaters of soil, detritus, leaves, dry grass, and just about any other form of cellulose.

Insect Skyscrapers

The morning was cold and the sun had just started to lighten the African sky behind a distant line of acacia trees, when a termite researcher named Gregor Schuurman went to work in Botswana's Okavango Delta. Termite mounds, built by the species *Macrotermes michaelseni*, stood all around the grassy floodplain like weird druidic monuments. Each one rose on a broad pyramidal base, and then narrowed into a finger of clay, slightly bent, as if beckoning to the sky. Schuurman, a doctoral candidate from the University of Minnesota, also looked a little druidic: tall, thin, with a goatee, and hair pulled back in a braided ponytail. He selected a 7-foot-tall termite mound, about average in this part of the delta (though 25-footers also sometimes occur). Then he swung a pick with a termite-nibbled handle and began to cut out a pit at the base of the mound. A half hour later, the clean slice of a shovel blade exposed the first signs of the termite world within: sculpted tunnels, which the termites had carved out mouthful by mouthful, cementing the round walls into place with their saliva.

"They've got a network of tunnels radiating from this mound for 160 feet in any direction," said Schuurman, "which means they're covering an area of 80,000 square feet with a single mound," roughly the equivalent of two football fields. Aardwolves, aardvarks, pangolins, and a few dozen other species would gladly eat the termite occupants, given the chance, but the tunnels enabled the workers to reach every potential food source in their vast territory without ever traveling more than about 5 inches on the perilous surface. "They're blind. They're soft-bodied. They dry out," said Schuurman. "They're very vulnerable creatures." And yet they are master builders. If termites were the size of humans, according to an estimate by the entomologist P. E. Howse, the largest *Macrotermes* mounds would be 5 miles wide at the base and a mile high, more than triple the height of our greatest skyscraper. It is an extraordinary effort, especially considering that the payoff hereabouts consists entirely of dead grass.

In the middle of the dry season, the grass was yellow and crackly underfoot. "You have to eat an awful lot of this stuff to get enough nitrogen to build tissue," said Schuurman. "So what this branch of termites has done is move the digestion outside their bodies. They use a fungus to help break down this grass, and it basically burns off a lot of the carbon, which comes out in the form of carbon dioxide."

Schuurman's digging sent up clouds of dust as the morning lengthened and turned hot. Up to now, the termites in this mound had mostly been fleeing well ahead of the calamitous intrusion of his pick and shovel. But gradually he exposed a great chamber under the center of the mound, and it was as if he had burst into the central assembly plant of the demented genius in some James Bond movie. Pale quarter-inch-long workers bustled everywhere. Soldiers a half-inch long, with big copper-colored heads and amber abdomens, charged out to defend the nest, snapping their curved mandibles to pinch the skin and drive off intruders.

Despite the defenders, Schuurman exposed the upper part of the central chamber, a half-dozen clay shelves, where orange, fibrous structures like honeycombs stood on stubby little feet. Having consumed grass at the perimeter of their territory, the worker termites come back and deposit fecal pellets on these combs, where the fungus does the work of digestion. Then the termites eat the fungus and their broken-down fecal pellets.

The chamber was a garden without sunlight. The whole purpose of the mound, Schuurman explained, is to create proper growing conditions for the fungus. Inside, it's like a rain forest in the middle of a savanna, 90 degrees Fahrenheit (32 degrees Celsius) even in winter, and 90 percent humidity even in the dry season. A complex system of chimneys runs up through the termite mound. (With a little modification, Schuurman once converted an abandoned mound into "the world's best pizza oven.")

Until recently, researchers believed these chimneys enabled the mound to function mainly as a cooling tower, dissipating heat produced as the fungus digests cellulose and burns off carbon. But research by Scott Turner, a physiologist at the State University of New York, suggests that the mound plays only a minor role in regulating temperature. Turner discovered that air moves both up and down the chimneys with a kind of tidal motion, depending on wind and external temperature. This draws oxygen in through the surface of the mound and expels carbon, exchanging vital gases much as do our own lungs. Altogether, the mound, the social system that built it, and the symbiotic fungus enable termites to dominate their world: They eat more grass than all the wildebeests, Cape buffalo, and other savanna mammals put together.

Even Schuurman's violent excavation would not threaten their dominance. Aardvarks routinely take minutes to do the kind of damage he had done in two hours of digging. And the termite colonies routinely recover. Moreover, Schuurman had not even come close to the real heart of the colony: the royal chamber, where the king and queen mate and produce their almost endless offspring, at a rate of 30,000 eggs a day.

The royal chamber, said Schuurman, is "about the size of a pack of cards, which doesn't give the queen much room to move around." Not that she does much moving, anyway. To fulfill her reproductive duties, the queen has swollen to about the size of a human index finger. In *The Soul of the White Ant*, Eugene Marais described her as "an unsightly wormlike bag of adiposity." The king maintains his youthful trim, never growing more than a half-inch long, but Marais happily added that he nonetheless remains passionately faithful to his bride. The two of them pursue their marriage in a chamber about 20 inches below the ground, which Schuurman described as "a very thick lump of clayed earth, which has very few tunnels through it. It's a real *sanctum sanctorum*."

The queen may live for 15 to 20 years and produce 250 million eggs. But her colony may survive her. When she dies, the pheromone that has kept most of her offspring sexually immature vanishes. One or more of her offspring may now become heirs to her throne.

When a colony's long dynasty eventually ends, a new termite colony may take over the abandoned mound. Or it may become home for other animals, from black mambas to wild dogs, and slowly fade back into the earth. But the earth itself is altered by the passing of each termite mound. Because the mounds are rich in minerals and organic material, plants rapidly colonize them, sometimes forming raised islands of trees in the grasslands. These islands may eventually grow together into continuous forest. Thus termites do not merely take forests down, piece by piece; they also build them up.

Happily Ever After

Back in New Orleans, I'd visited a building called the Upper Pontalba, a National Historic Landmark, three stories high and a block long, with lacy wrought-iron balconies. City officials had been boasting that here, in the heart of the war on Formosan termites, they had for once succeeded in exterminating the invader. The city's newfangled poisons had actually managed to kill a colony on the ground floor and in the trunk of an old tree out on Jackson Square, 300 feet away. But that night on the third floor, in one corner of a fine old apartment, we spotted the telltale signs of infestation. Within the walls, the termites had built satellites called mudpacks, where this year's alates were preparing to launch themselves for their big mating flight. The termite population had metastasized.

Termites have spent more than 100 million years refining their business of covertly feeding on wood, and it is hard to be too optimistic about our ability to stop them. Back home in Connecticut, I checked in on the progress of the house I was building, letting my house-proud eye play over all the handsome wooden details. And then I recalled something I'd heard from a New Orleans schoolteacher, after a crew had just laid down a new wood floor in her termite-riddled classroom.

"Thank you," she said, on behalf of the tireless termite world beneath her feet. "You just gave 'em free lunch."

The Monkey Mind

*I*t was midmorning at the Jackalberry Café, a stand of big, branchy trees on an island in Botswana's Okavango Delta. Dorothy Cheney and Robert Seyfarth, a husband-and-wife research team, were surrounded by baboons, many of them foaming at the mouth. With their long muzzles and sinister copper eyes, the baboons looked a bit like mad dogs. But the foaming was just a side effect of eating jackalberries, which taste like chalk cut with lemon juice. The baboons had their cheeks crammed full, and their steady gnawing was about as threatening as the grinding of a pepper mill. They took no notice of the human researchers wandering among them with notebooks and a microphone.

A fight erupted in the trees overhead, and Cheney and Seyfarth started to sort out the antagonists. A high-ranking juvenile named Puncture had it in for a cousin named Palm. The two of them leaped from branch to branch, squealing and searching for allies. Puncture unwisely tried to enlist another cousin named Plimsoll. Everyone ended up on the ground, where Plimsoll groomed Palm, who happened to be his sister, the two of them feigning utter indifference to Puncture. Puncture slapped the ground in front of them and popped big-eyed threats in Palm's face. But Palm just lolled her head to one side and closed her eyes, deep in the bliss of being groomed. Puncture grabbed a branch and shook it under Palm's nose. No response. It was infuriating.

"This is like a really bad Thanksgiving dinner," said Seyfarth. "Sometimes the little cousins fight."

"Sometimes everybody fights," said Cheney.

The easy comparison to humans—anthropomorphism, that is—was odd. Other researchers have built their careers on the popular premise that monkeys and apes are almost human. Cheney and Seyfarth take the opposite view. Over the past 25 years, their ingenious experiments with vervet monkeys and baboons have sometimes revealed extraordinary richness in the monkey mind, particularly when it comes to social knowledge about one another. But more often, Cheney and Seyfarth have discovered severe limitations on intelligence and communication in monkeys. "They're not furry little humans," Seyfarth said now. "They're just monkeys."

This affront to our sentimental notions about the natural world may be a little easier to take in the context of baboons, which are ugly, unendangered, and politically incorrect. "Baboon" has become synonymous with boorish male behavior, and with good reason. Males are twice the size of females and tend to do whatever they want, whenever they want, often in an egregiously offensive manner. Baboons can also be destructive to human interests. Farmers regard them as crop-raiding, lamb-killing vermin, so much so that in 1999 a South African development group announced plans (quickly abandoned in the face of international outrage) to finance a slaughterhouse specializing in baboon salami.

Still, baboons are primates, like us. We had a common ancestor roughly 30 million years ago, when the apes (now represented by chimpanzees, bonobos, gorillas, orangutans, and humans) began to diverge from the Old World monkeys (now represented by baboons, macaques, vervets, and many others). Humans are, of course, much more closely related to chimpanzees; our two species diverged from a common ancestor just 5 or 6 million years ago, and chimpanzees share about 98 percent of our genetic material. But the figure for baboons is also high, about 92 percent. And yet Cheney and Seyfarth's work is a reminder that a genetic difference of just a few percentage points can translate into vast, unbridgeable gaps between species.

The Okavango Delta, where the pair do their work, is a wilderness of open grasslands and scrubby dry forest, broken up into islands by seasonal flooding. Baboon Camp, on the edge of one such island, consists of a few tents and a reed-roofed kitchen with freezer, refrigerator, a feeble gas range, and an old dugout canoe hull for a countertop. Hippos sometimes graze among the tents, and when wild dogs hunt, panicky impalas can come bowling down a trail without warning. Postdoctoral students work here during the school year. Cheney and Seyfarth, who teach at the University of Pennsylvania, come out during summer breaks, with their teenage daughters Keena and Lucy.

Keeping Company with Troop C

To find the baboons known as Troop C, the researchers walk out from camp each morning, ears alert for Cape buffalo, elephants, and lions. They carry no weapons, so avoidance is their only defense, and tree climbing a last resort. Water crossings can be particularly hazardous, because crocodiles have a nasty habit of launching themselves missilelike from the bottom. The baboons of Troop C are also aware of these dangers and tend to stick to a home range of less than 3 square miles. They sometimes turn up in the camp itself, where, as Keena recalled with dismay, they occasionally pluck a roosting fruit bat from a fig tree and eat it headfirst "like an ice-cream cone."

The research team can identify all 84 individuals in Troop C, usually by physical features like Gary's crooked tail or Roy's horsey muzzle, and sometimes by sound, as when Shashe gives her warbling, operatic lost call. They also know each animal by personality, a factor scientists cannot quantify, but feel deeply. Seyfarth once mentioned a domineering female named Sylvia to a predecessor at Baboon Camp, who promptly replied, "That *bitch*! What's she up to now?"

Cheney and Seyfarth use the monkeys' own vocalizations to determine what the monkeys themselves know, and it appears that the baboons also recognize one another by voice, rank, family, and perhaps even personality. The researchers' technique is to record a known individual giving a reconciliation grunt, a fear bark, an alarm call, or some other meaningful vocalization. They play back the call to find out how other individuals in the group respond.

In a typical experiment involving the booming *wa-hoo* calls sounded by rival adult males, Cheney positioned herself to videotape a female while Seyfarth maneuvered a 4-foot-high speaker in a wheelbarrow behind a bush. Then, on a signal from Cheney, he played the call. Cheney recorded the female for the next 15 minutes to see if the vocalization altered her behavior.

Such painstaking work has produced some intriguing results. When Cheney and Seyfarth were starting out in the 1970s, scientists generally believed that monkeys could communicate little more than distress. Using playbacks, the pair showed that vervet monkeys were in fact capable of making subtle distinctions. They rarely sounded an alarm call when they spotted a snake eagle but almost always did when they saw a far more dangerous martial eagle. Sometimes they varied the rate and intensity of their calls depending on whether relatives, potential mates, or high-ranking rivals were within earshot.

Cheney and Seyfarth also found tantalizing evidence that vocalizations may sometimes work as abstract representations of ideas. For example, vervets learned to ignore a starling's eagle alarm call when it was played repeatedly without an eagle present. Then Cheney and Seyfarth played the vervets' own eagle alarm, and the vervets ignored that too. Knowing one call was a false alarm meant knowing its "synonym" was false as well. But if Cheney and Seyfarth then played a starling's leopard alarm call, everybody scrambled for cover.

The Social Order

One morning out at a place called the Palm, not far from the Jackalberry Café, a couple of baboons were sitting around taking in the sun, being groomed by subordinates, and waiting for breakfast to fall out of the sky; another baboon, eating palm nuts at the top of a tree, now and then inadvertently shook a few to the ground. Power and Selo, the king and queen of Troop C, took the first couple of nuts. When they moved off, Sonny, the number two male, slipped into their place. Sonny threatened Gary with a sharp lifting of the eyebrows. Gary immediately saved face by passing the threat on down the line to a hapless juvenile. Watching the males jockey for position, it was hard to avoid the impression that baboon social life is largely about what kind of trouble the big boys are up to now.

In fact, females are the stable element in baboon society. They stay with the troop throughout their 20- or 30-year life spans, and mothers tend to pass on social status to their daughters. For instance, Sylvia and Selo, two sisters with beady eyes and imperious

Doom

*T*he first time I visited Africa was early in the AIDS epidemic, and every time I woke up on the long drive from the airport, the sides of the road seemed to be lined with merchants selling coffins in the open air and billboards advertising DOOM. Maybe I was imagining it, because I have never seen anything like it since, and on such a messy, sprawling, difficult continent, a newcomer naturally suffers vaporous feelings of dread.

Later, I was having dinner at my hotel in the foothills of the Rwenzori Mountains, which I was planning to climb. As a lifelong lowlander, the prospect made me nervous. I was thinking about another climber who had died there just a few months before, of altitude sickness. Then a woman on the housekeeping staff came up to my table and said, "Please, may I have your key, sir? I need to spray your room."

"What are you going to spray it with?" I asked.

"Doom," she said sweetly. I must have looked stricken, because after a moment, she added, "For mosquitoes." It was an insecticide.

This has been the pattern for all my travels since then, in a half-dozen African countries—paranoid anticipation giving way to (let's not go overboard) cautious enjoyment. The people I have met have been almost unfailingly kind and generous, even in the face of ugliness and stupidity.

There are of course rules: Take your malaria meds. Steer clear of Cape buffalo. Do not stop by the side of the road at night even in remote regions of some countries; people will appear and, coming from the developed world, you will have more money than they can imagine. And I suppose I should add, "Don't drink the water," a rule I broke in spectacular fashion on that first trip.

Explorers like Richard Burton and Henry Stanley had once sought the Rwenzori Mountains as Ptolemy's legendary "Mountains of the Moon," and in nineteenth-century fashion, I somehow ended up with two Bakonjo guides and a line of eight porters, who vied for the chance to carry 48 pounds apiece up steep slopes at high elevations in rain and snow for $4 a day. (I didn't actually have that much baggage. So maybe they were carrying party supplies. We were also followed up the mountain by several baby goats who did not realize that coming back down was not on their agenda.)

The Uganda name for these mountains, Rwenzori, means "rainmaker," or "hill of rain," or maybe even "the great leaf in which the clouds are boiled," and scores of streams come rushing down the flanks. We followed the river valleys up, with ferns and creepers reaching in from both sides of the trail to brush us like beggars in a crowded street. Heather plants grew as big as trees, their branches wrapped in sleeves of moss up to 3 feet thick, from which ferns and orchids also sprouted. Elephants and buffalo used to roam the foothills, but they got wiped out by heavy poaching in the 1980s. And above 8,000 feet, the guides assured me that it was safe to drink directly from the stream. With the wildlife gone, there was no risk of giardia or other contaminants.

The Bakonjo were dependable. Five days into the trip, on the final leg up to the summit at almost 17,000 feet, they waited patiently as I paused to listen to the cross-rhythm of my roaring lungs and racing heart. We were roped together on the glacier, and snow goggles turned everything green, but the disparity between us was still striking. I was wearing expensive leather climbing boots and thick Gore-Tex gloves; they had rubber gumboots and bare hands. A sleeve of one of their tattered jackets looked as if had been amputated and reattached. But they were fine, and I was struggling. "You can make it," one of them assured me. Then on the slope just below the summit, I lost my footing and went skittering down toward a steep drop-off into the Democratic Republic of the Congo, a destination the U.S. State Department was advising Americans not to visit, especially by free fall. My ice axe cut a plume in the snow, finding no purchase. But then the rope tightened, and my guides gently hauled me back up to safety.

They seemed to be right about drinking the water, too. In the thin air at high altitude, I lost my appetite for food, but otherwise stayed healthy. For part of the climb I traveled with a group of British doctors, who happily advised me that a warlord just across the border in Rwanda was offering $1,000 for the head of an American. Then they sang "The Star-Spangled Banner." I felt safer with the Bakonjo.

The trip back down was long and difficult, in a steady rain, through mud with the consistency of watered-down chocolate icing. Near the end, I came to a clear stream, threw myself down without thinking, and buried my face in the water to drink. Then it dawned on me that I was well below 8,000 feet. I looked up and turned my head sideways. There, grazing just 50 yards upstream, was a herd of cattle. Doom, I thought.

But of course it wasn't doom at all (just a week of violent gastrointestinal turmoil). If I had been humbled at times by my own ineptitude during the 10-day trek, I'd also felt exalted at times by this trip into the voluptuous heart of the mountains. With the help of the Bakonjo, I had done something I had been afraid to do, something I was not sure I was even capable of doing. My guide, Ezra Mutahinga, sent me away with a Bakonjo proverb: "Fearing," he said, "is not dying."

leonine strides, belong to a female line that has dominated Troop C for more than 20 years.

Males, on the other hand, move out into neighboring troops when they reach maturity. There, if a newcomer has promise, he may eventually challenge the local alpha male with *wa-hoo* contests and physical force. The new alpha does his best to kill any unweaned infants in the troop, as a way of bringing the mothers back into reproductive condition so that future infants will be his. The alpha also tries to monopolize mating with any female who comes into season. Cheney at one point summed up the paths of male glory this way: "Run wild, kill babies, you're there."

But the social lives of baboons are more subtle than that. Once you know the individuals and their family histories, the dull round of jackalberry and palm nut feasting opens up like a plotline for a novel. Or, as Cheney puts it: "The rules with baboons are the same as in a Jane Austen novel—maintain close ties with your relatives and try to get in with high-ranking animals."

For the moment, Troop C was a dysfunctional family, even by baboon standards. The baboon named Power had somehow managed to become the alpha male without ever leaving home, and he didn't know quite how to behave. When Selo developed the swollen red rump that advertises sexual fertility, Power hung out with her. But nothing much happened. "He's only got half the story," said Cheney. The two of them sat around together chewing palm nuts and wondering, perhaps, if that's all there was to love. Like humans, baboons have inhibitions against incest; growing up together had apparently chilled any romantic chemistry between them. Cheney and Seyfarth were hoping Sonny, an outsider, would work up the nerve to overthrow Power and make things a bit more interesting. "Come on, Sonny, bust a move," their daughter Keena urged.

"It's like sitting in an Italian café nursing a glass of wine, and seeing the teenagers and the young adults coming and going," said Seyfarth, "each with a complex story." Sometimes someone slipped away to avoid running into a more dominant rival. Sometimes two friends met and strengthened their bonds with grooming.

But were the animals themselves aware of the unfolding plotline? Apparently so. When Cheney and Seyfarth played back a baboon's vocalizations, the individual's identity unmistakably influenced how others responded. An alpha elicited a submissive grunt from a subordinate. A cry for help caught the attention of a sibling. The baboons' knowledge about who's who in the troop, and about the workings of rank and family relationships, added up to what the researchers call "laser-beam intelligence." As Seyfarth put it: "Social intelligence is for primates what celestial navigation is for arctic terns."

What's on Your Mind?

Still, laser-beam intelligence, for all its sharpness, is by definition narrowly focused. Outside the social sphere, baboons didn't seem to know much, even about matters of survival. Cheney and Seyfarth once arrived at a study site just after a pride of lions had killed a Cape buffalo. The bloody, dismembered carcass was lying in plain view. But the baboons strolled past, apparently not recognizing that this crime scene was a sure sign of lions. They became alarmed only when they actually spotted the lions resting in the brush nearby. Likewise, neither baboons nor vervet monkeys seemed to recognize that an antelope carcass stashed 20 feet up a tree meant that a leopard, their worst enemy, was in the neighborhood.

What's even more shocking, since social intelligence is their specialty, is how little the baboons seemed to know about one another's minds. One morning, after a languid session of sunbathing at the Termite Mound Spa, the troop moved off in desultory fashion. As they spread out through the woods and randomly foraged, baboons on one side or the other gave "contact" or "lost" calls. For humans, it was natural to assume that the baboons were exchanging barks as a way of signaling, "Hey, I'm over here. Where are you?"

But when Cheney and Seyfarth tested this idea using playbacks, the baboons almost never barked in reply—unless they happened to be lost themselves. Even mothers failed to bark back to their offspring. It was as if the monkeys did not realize they could use vocalizations to inform or influence the beliefs of their fellow monkeys. They barked not to say, "Hey, we're over this way," but merely to lament their own sorry state of being lost. And it only worked as a contact call because, in the course of any move, several baboons on different sides of the troop tended to announce that they were lost at about the same time.

Cheney and Seyfarth have gradually come to the conclusion that monkeys don't actually recognize that other monkeys have minds. They feel grief themselves, for instance, but almost never comfort other monkeys who happen to be grieving. They do not seem able to put themselves in another monkey's place. Sylvia, for instance, once made a long water crossing with her baby clinging to her belly. Since Sylvia herself could breathe, it did not dawn on her that her submerged baby couldn't. So it drowned at her breast.

The absence of what academics call "a theory of mind" may help explain why monkeys and apes (with the exception of humans) have never developed a true language: they don't invent words because doing so would require both speaker and listener to attribute intentions and beliefs to one another.

Apes in the laboratory have sometimes learned to understand scores of words. With intensive training, a few animals, such as the chimpanzee Washoe and the bonobo Kanzi, have become famous for their ability to respond to human language in surprisingly complex ways. But the circumstances were completely artificial. "You can teach a bear to ride a bicycle in the circus," said Seyfarth, "but it doesn't tell you much about what bears learn to do in the wild." And even in the laboratory, no animal has attained anything like true language.

Humans, on the other hand, embody theory of mind in wild

excess. We know what we know, and we know that we know it. We possess the playful, curious, strange, and sympathetic entity called human consciousness. Beginning sometime before age four, as the growing mind leaps ahead, we start to intuit what our fellow creatures know too. And one result is that we can hardly keep ourselves from projecting our own thoughts and feelings not just onto one another but also onto other animals and even objects. "It's kind of like my mother with an ATM machine," said Seyfarth. "She assumes there's a person behind it. She's not prepared to deal with the fact that it's just a machine."

The animals we anthropomorphize often behave as we'd expect a creature with thoughts and emotions to behave. They may in fact have thoughts and emotions. But there is no evidence, in Cheney and Seyfarth's view, that they can project either outside themselves, as humans routinely do. It's a fine point that the anthropomorphizing human mind is ill-equipped (or unwilling) to grasp. "When we talk about theory of mind," said Seyfarth, "people eventually say, 'OK, we'll accept that monkeys and chimps are not like people. But my dog . . .'"

Back at camp one day, Cheney and Seyfarth joined in eagerly as their daughters discussed their own dog, a corgi named Eliot. Eliot likes to sleep under Lucy's bed, they agreed. But then he starts to feel guilty, and worries that he ought to be protecting the whole family. So he creeps out and spends the rest of the night on the landing. The two scientists went along with this anthropomorphism, though they knew it had much more to do with what was going on in their daughters' minds than in their dog's. Projecting our own feelings onto animals is human nature—perhaps even human nature at its most appealing.

A little later, in the camp kitchen, Lucy had set aside *Harry Potter* for an attempt at *Cry, the Beloved Country*, Keena was deciphering *War and Peace*, and their parents were logging field notes on a laptop. Out on the plain, Shashe began to give her lost call. It was tremulous and pathetic, and after a while you wanted to go out and lead the poor creature by the hand back to her troop. "Shashe's saying, *'Ple-e-e-ease*, somebody answer,'" Cheney remarked. But the baboons in the bush were mere monkeys, and though Shashe's calling went on for half an hour, not a single soul replied.

Hummers

*I*t wasn't quite six on a radiant Arizona morning when Marion Paton, a retired school-cafeteria manager with big tinted glasses and golden hair, padded into her kitchen and glanced out the picture window. Nineteen people sitting in her backyard stared back through their binoculars. There was a fat, bearded man in a luau shirt and a hat with the brim bent up in front. There was a slight, older man, mild as a parson, with round, wire-rimmed eyeglasses and a blue zip-up jacket. There was a woman in clamdigger slacks, white socks, and a bush hat studded with birding pins. The assembled crowd was engaged in the behavior Paton calls "whooping and dooping."

"There's a very nice violet-crowned on number four," said a guide, "and you can really see that white breast. *Oooh, oooh*, it's that male rufous. Flash that tail at us again!"

Did Paton perhaps long for a little privacy, at least until the coffee dripped? The question astounded her. "What for?" she asked. She did not even mind the time a couple of duck blinds turned up in the yard, with television cameras trained on her house. "I love people," she said. "I love nature."

She wandered outside. "Canyon towhee running around by the hose," said a guide, and the binoculars whipped right. "Yellow warbler singing behind us." The crowd swiveled around. Then, one by one, the lenses came yearningly back to Paton's modest house, where feeders full of sugar water hung from the eaves. They were swarming with hummingbirds, which John James Audubon once likened to "a glittering fragment of the rainbow."

Raging Appetites

In fact, the tail feathers and metallic gorgets coruscating in the moist air looked more like the whole damned spectrum. But the action here was too intense for rainbow analogies: broad-taileds, Costas, and calliopes jostled around every feeding hole, and other hummingbirds twittered impatiently nearby. Paton's house in the remote town of Patagonia happens to be in the middle of a major north-south flyway. It's also right next to a nature preserve. On a good day, she gets 11 hummingbird species (of the 17 recorded in southeastern Arizona) and maybe 50 bird-watchers. Down the road toward Nogales, Jesse Hendrix, the keeper of another unofficial hummingbird way station, boasts 10,000 hummers a day on his 150 feeders at the height of the migration, and he can run through 150 pounds of sugar a week. Paton prefers to keep things modest, with no more than eight feeders.

"These creatures have a following like mythical beasts," said one of the guides, a little ruefully. "There are people who don't care anything about birds, or other wildlife or nature, but they love hummingbirds. We had one woman tell us: 'I just love hummingbirds and unicorns.' And I don't think she drew a distinction between the two."

The guide's name was Tom Wood. He was from the Southeastern Arizona Bird Observatory. "People come in," he continued, "and they say, 'They're so tiny, and they're so sweet,' and we'll say, 'Well, they *are* tiny.'" Wood trained his binoculars on a feeder. The glittering fragments of rainbow were at that moment swatting and screaming at one another in a relentless bid to get to the head of the line. "They're fighter pilots in small bodies. We've seen a bird knock another hummingbird out of the air and stab it with its bill. People still don't believe it. They think they're little fairies." He shrugged. "We're probably lucky these things aren't the size of ravens, or it would not be safe to walk in the woods."

Hummingbirds are among the smallest warm-blooded animals on Earth, and though it may be heresy to say so, they are also among the meanest. The bee hummingbird, a Cuban species, weighs less than a dime, and even middleweight species like the rufous and broad-tailed hummingbirds weigh less than a nickel. Their size makes them cute—and also dictates their fretful, bickering, high-rev way of life. A big stolid raven can store enough energy to plod through good times and bad. But even in the best of times, a hummingbird is a slave to its raging metabolism.

Hummingbirds get most of their energy by sipping nectar from flowers, and a typical hummingbird needs 7 to 12 calories of energy a day. This sounds idyllic, until you do the math: it's the equivalent of a 180-pound human having to scrounge up 204,300 calories a day, or about 171 pounds of hamburger. To keep itself alive, a hummingbird must manage to find as many as 1,000 flowers and drink almost twice its weight in nectar daily. It's enough to give even a very pretty little bird the personality of a junkyard dog, not to mention an urgent need to pee. A scientific paper about the rufous hummingbird includes this endearing notation: "SOCIAL BEHAVIOR: None. Individual survival seems only concern."

Life in the Fast Lane

One morning at the Rocky Mountain Biological Laboratory, in the village of Gothic, Colorado, a researcher named Bill Calder reached into a hardware-cloth cage and gently folded a fresh-caught rufous in the palm of one hand. "Hello, little man," he said, and to a visitor, he added, "He's straight from God-knows-where. Somewhere between Montana and Alaska." It was only mid-July, but this thumb-sized bird had already flown roughly 2,000 miles north from his winter home in Mexico. With a combination of luck and cussedness, he had established a territory, defended it from rival males, put on courtship displays for passing females, and mated with as many of them as possible. Now he was on the road south again.

"You've got to see this in the sun," Calder said. He held the bird up, and the patch of color at its throat glowed with deep, shifting reds and oranges. The throat patch, or gorget, is likely a display device for attracting females and intimidating rivals, but the iridescent colors were entirely appropriate to the rufous hummingbird's high-energy way of life. "It's like a burning coal," Calder said.

He took the bird inside, weighed it, crimped a tiny metal identification band on one leg, returned outdoors, and set it free. He and his wife, Lorene, had been doing this for 30 summers at Gothic, to figure out the consequences and requirements of body size in different birds. Calder, a professor of biology at the University of Arizona, was lean and energetic, with a gray Vandyke beard, a beaky nose, and thin lips framed by deep furrows arching down from the corners of the nose. He tended to flit from subject to subject, and he was relentless about his work. The Calders' cabin at Gothic was 9,500 feet above sea level, and even in July strips of snow in the high draws still sent rivulets and waterfalls down the gray flanks of the Elk Mountains. It was a good location to study hummingbirds, although as a genus they are basically tropical. Hummingbirds are a New World family (the sunbirds of Asia and Africa are only superficially similar), and more than half the roughly 320 species live near the equator.

But hummingbirds are also opportunistic. They will fearlessly investigate any potential food source. This is how Calder once got french-kissed by a hummingbird: "I was out there working with my mouth open—I was always a mouth-breather—and a hummingbird flew up to me and put her bill in, and I actually felt her tongue on my tongue."

Hummingbirds also routinely explore new habitats, because the cushier habitats quickly fill up with other hummingbirds. Thus some equatorial species have evolved to live high up in the Andes, and other species migrate 500 miles across the Gulf of Mexico, or down the Rocky Mountains from Alaska to Mexico. But why travel the colder mountains instead of, say, the plains? Calder indicated the swaths of yellow, violet, and red wildflowers, which lay like veils across the green foothills above Gothic: "Abundant flowers in a compressed growing season," he said.

In fact, hummingbirds seemed to be everywhere on the slopes around the Calders' cabin. Broad-tail males perched on high branches and electric wires, each fiercely guarding a feeder or a patch of flowers. One of Calder's students found that a male broad-tail at a feeder typically flies more than 40 sorties an hour to drive off rivals. Roughly another 45 times an hour, he shoots 60 feet straight up in the air and back down in a gaudy courtship power dive, his wingtips giving off a metallic trill, urgent as a bicycle bell. This is incredibly taxing. A hummingbird's heart beats more than 1,200 times a minute in flight; his wings hum at 2,280 revolutions per minute.

Yet, even with a patch of succulent flowers at his feet, a male broad-tail actually eats very little for much of the day. A full belly would give him the aerodynamics of a lumbering old bomber, reducing his ability to chase rivals and display for females. So he waits till dusk, and then goes on a 20-minute binge, hitting flower after flower until his crop sags and his weight balloons by a third with the fuel he needs to survive the frigid night. Then he abandons his territory and, contrary to our expectations, flies uphill. Thermal inversions make the mountains about 5 degrees Celsius (or 9 degrees Fahrenheit) warmer at night than the valleys, and the males know where the warmth is. Given the speed with which small bodies lose heat (think of a spoonful of soup versus a full bowl), such subtle adaptations can save a bird's life.

The females, meanwhile, tough it out in the valleys with their offspring. But they've also evolved tricks for surviving harsh mountain nights. In a stand of evergreens and aspen not far from his cabin, Calder pointed out a female broad-tail's nest on a branch overhung by a higher bough in a spruce tree. The bough, he said, serves as a roof over the nest, reducing nightly heat loss. The nest itself was about the size of a baby's fist, flecked with bits of green lichen for camouflage. The female weaves it together with spider webbing. Hummingbirds may also insulate their nests with down or the feathery white aspen seeds that sometimes drift through the valley like a snow flurry. But the female's last resort against cold is torpor: if she doesn't have enough energy to get through the night, she can turn down her thermostat, cutting her body temperature in half. In torpor, her metabolism slows two to three times for a 10-degree-Celsius (or 18-degree-Fahrenheit) drop in temperature.

The female at this particular nest was a small mousy broad-tail with a mottled green back, and she welcomed her human intruders by perching on a nearby branch, twitching her head from side to side, and screaming *chip-chip-chip*. It made her visi-

tors feel faintly ashamed, like being scolded by a mother for waking the baby.

Calder backed off, and the broad-tail returned to the spruce
tree in stages, pausing in midair and glancing around, as if to
make sure she was not being followed. She began to feed, hovering in the air around the nest with her head cocked, then zipping
up, down, left, and right, seizing gnats with her bill. Despite their
reputation as nectar-sippers, hummingbirds routinely get their
fats and proteins by eating insects. When she'd filled her crop,
she perched on the rim of the nest, and her two nestlings rose up
in a flurry of soft feathers and gaping maws.

What happened next was appalling: the mother stuck her bill
halfway down one nestling's throat, as if she'd mistaken her baby
for a sword-swallower. Then she started jabbing up and down like
a sewing machine. "That much bill and that little chick," Calder
mused. "I'm always afraid she'll overshoot." She was regurgitat-
ing food and literally packing it in, to get her youngster to eight
times its hatch weight in just two weeks. When she was done, she
swooped away from the nest, *chip-chip-chipped* some more, and
hovered close enough that her human visitors could feel the turbulence from her angry wingbeats on their faces.

Wicking and Licking

The ancestors of hummingbirds probably started out feeding on
tiny insects around flowers, and only incidentally got their noses
into the nectar. But they took to flower-feeding like a small child
to lollipops. Hummingbirds and certain flowers have subsequently adapted in all kinds of weird ways for the blissful moment
or two when they come together. In the Andes, for instance, certain passionflowers have developed an elongated tubelike shape.
Local hummingbirds in turn have evolved 4-inch-long sword-

bills for reaching deep down to the nectar at the bottom of the tube. When they finish feeding, the hummingbirds inadvertently carry a dusting of pollen on their bills and heads and deliver it to fertilize other passionflowers, sometimes miles away. The birds are the flowers' primary pollinators.

Hummingbirds have also evolved incredibly long, specialized tongues. If you hold a hummingbird in your hand and offer it a feeder, you can sometimes see the white flickering of the tongue entering the feeder hole just ahead of the bill. If you blow gently on the bird's head feathers as it feeds, you can actually see the tongue muscles pulsing under the translucent flesh at the back of the skull.

This may seem anatomically unorthodox. The tongue itself fills the hummingbird's bill, so the muscles that support the tongue actually run back around the spinal cord, up the outside of the skull and over the top, to be anchored between the eyes. In some species, the tongue is fringed along the outer edge, which may help entangle insects. In some, the tongue ends in two troughs with which the bird draws up nectar—not by sucking but by capillary action—as the tongue flicks in and out. While the human spectators are whooping and dooping, the hummingbirds, in Calder's words, are wicking and licking.

So far, this is pretty straightforward. But as the relationship between flowers and hummingbirds evolved, certain mites figured out how to get in on the party. These tiny relatives of ticks and spiders eat nectar and pollen, and each mite species has evolved to feed on particular species of flowers. Getting from one flower to another can be a problem—especially given that mites are blind and may specialize on a bromeliad 100 feet up a tree in the middle of a rain forest. So the mites have evolved to use hummingbirds as their C-47s.

Sooner or later a hummingbird will show up to feed on the

flower a mite has been busy plundering. Then, according to University of Connecticut biologist Robert Colwell, the mite sprints up the hummingbird's bill and hides in its nasal cavity. The hummingbird doesn't seem to notice the stowaway. Colwell has found as many as 10 or 15 mites per bird, and because a hummingbird typically feeds on many different flower types, the mites have at times belonged to as many as five species.

The mites are only along for the ride to the next preferred flower. Colwell describes them as "perpetual airline passengers that carry out all their mating and feeding in airport lounges." They apparently decide which flower is Gate B7 by the floral scent sucked in four times a second on the hummingbird's breath. Getting off at the wrong flower can mean death for the mite. But it must make its decision instantly and move quickly, because hummingbirds spend only a few seconds at any one flower.

Almost everything else about hummingbirds is adapted to the flower-loving life, most notably their method of flight. They need to stop and hover precisely enough to extract nectar from each flower. Relative to their size, they have the largest flight muscles of any bird, up to 30 percent of their total weight, anchored to a keel-shaped sternum. They've also got twice as much heart as might be expected for their size, and a denser concentration of red blood cells for better oxygen storage. Their wings are short, and the bone structure is nearly all hand; they have an extremely short humerus, ulna, and radius, the equivalents of our upper and forearm. But unlike other birds, hummingbirds can rotate their wings in a figure eight—much as our wrist enables us to rotate a hand—due to their remarkably flexible shoulder joints.

When a hummingbird is hovering, its wings flap through the air horizontally, rather than up and down. As the wing sweeps toward the front, the leading edge rotates forward, for lift on the downstroke. Then, as the wing sweeps to the rear, the leading edge rotates back, for lift on the upstroke too. Other birds, with their up-and-down flapping, get lift only on the downstroke. But the hummingbird's flexibility enables it to hover, back away on the wing, and even fly nearly upside down. One reason hummingbirds sometimes drive off hawks and other birds a hundred times their size is that they can outmaneuver them.

Remembering Every Flower

Hummingbirds are also smart enough to know where to fly— and where not to. One evening outside the Nordic Inn in Mount Crested Butte, a few miles down the valley from his cabin, Calder drove long stakes into the ground and set up a mist net, which is like a portable spiderweb, 7 feet high and 36 feet long, screening off a couple of feeders hanging from the eaves of the hotel. The Nordic has been putting out these feeders for the past 30 years, and migrating hummingbirds have come to count on the service. In spring, the hotel is one of their base camps, for forays to test whether the wildflowers are blooming yet at higher elevations. If the owners forget to put out their feeders in time, the first hungry hummingbirds to arrive raise holy hell. It's a common experience among birders: hummingbirds not only get from Mexico to Alaska and back, but also appear to remember flower patches and feeders en route. One biologist complained that hummingbirds continued to show up at a regular feeder site for two years after she stopped putting out the feeder.

"All that memory in a brain case as big as Abe Lincoln's head on a penny," Calder said. But relative to body weight, the hummingbird's brain is actually bigger than ours, and hummingbirds are also shrewd enough to adjust to changing circumstances. As Calder talked, a broad-tail came whirring in and stopped in midair to contemplate the almost invisible mist net that now separated him from his feeder. Then he flew straight up, like a helicopter, over the top, and straight down to the feeder.

What's more remarkable is that hummingbirds actually remember individual flowers over the course of a day. They also seem to make decisions about when to revisit a flower based on how quickly it can replenish its nectar supply. One morning back in Arizona, Susan Wethington, a PhD student at the University of Arizona, set out a rack of 16 artificial flowers with differing rates of nectar replenishment. Some stayed empty, some refilled slowly, some quickly. A black-chinned, humming through on his second visit of the day, consistently hit the fast-replenishing flowers and skipped the empties. This sort of behavior has earned humming-birds a reputation, among scientists, as "nature's greatest efficiency experts."

Hummingbirds do not stick exclusively to red flowers. Their legendary fondness for the color may actually be one of the great myths about how animals divvy up a resource. According to the classic ecological explanation, flowers evolving for pollination by hummingbirds tended to shift toward the red end of the color spectrum, because bees and other pollinating insects can't see red very well. But Nick Waser, an ecologist from the University of California at Riverside, whose cabin in Gothic was just up the road from Calder's, dismissed this as "a great story to teach undergraduates."

Studies have shown that bees are capable of seeing red (although less well than other colors), and hummingbirds can learn to respond to any color. They are not interior decorators. What they want is nectar, they want it now, and they will concentrate on whatever color flower happens to be giving it to them at the moment. Moreover, hummingbirds will visit bee flowers, and vice versa. So why do hummingbirds turn up so often on the abundance of red flowers in the American West? One possible explanation, in Waser's view, is that red may simply stand out better against a green background.

Specialization, said Waser, isn't what got hummingbirds where they are today, or where they are going to be. (Anna's hummingbirds have lately expanded their breeding range from California into Arizona; rufous hummingbirds have launched an invasion of the Southeastern United States.) On the contrary, hummingbirds are the ultimate opportunists.

"They're just such ballsy birds," Sheri Williamson was saying one afternoon outside Sierra Vista, Arizona. "There's somebody in there. You look in some other bird's eyes and it's like looking in the eyes of a cow. But hummingbirds are so aware of what's going on."

Williamson, who runs the Southeastern Arizona Bird Observatory with her husband, Tom Wood, was conducting her weekly hummingbird banding session at the San Pedro Riparian National Conservation Area. They had a feeder set up, enclosed in a mist-net tunnel, which was closed off at one end. Wood, with a big gray beard and a contented red face, stood at a distance. When a hummingbird came to the feeder, he lumbered toward the entrance and tried to panic the bird into the netting. Sometimes it worked.

When Wood caught his first customer, a young black-chin female, he took it over to Williamson. After banding the bird, she wrapped it in a scrap of cloth "like a birdie burrito." The wrapping calmed the bird for its weighing and medical exam. Williamson blew on the feathers to examine the fat content beneath the tissue-thin skin. "She's already been in the first fight of her life," Williamson said, "because she's missing one of her tail feathers." Then, the examination finished, Williamson held the bird up to a feeder. "This is like juice and a cookie at the blood bank," Williamson said. "We're just trying to pay her back a bit."

When Williamson later held another hummingbird to the feeder, it guzzled sugar water till its crop swelled goitrously at the back of its neck. In a couple of minutes of feeding, its weight went up by a third. Williamson placed the bird in a visitor's hand to be released. It sat for a long time, unperturbed, its body thrumming, not with its heartbeat ("That's too fast for you to feel," Williamson said) but with its usual 250 breaths a minute.

The hummingbird blinked and looked around calmly with its glossy black eyes, lord of all it surveyed. Finally, it shuttled off in a rush, leaving behind, as a warm token of disregard, a droplet of urine in the hand that had cradled it. Williamson was ecstatic. "Aren't they just wonderful birds?" she inquired.

Tucking Ptarmigans Strictly Prohibited

Some time ago in the pages of a national magazine, I lampooned a Texas politician so eager for television coverage that he stuck his hand into a fire ant mound on camera and held it there while promising to whip the fire ant problem for good. Then another news crew showed up and the big buffoon did it all over again.

I was just having fun, beating up on our leading statesmen. It is what we high-powered media types do in lieu of working for a living. But then I wrote the line that was to come back and haunt me. It was about my own investigation of the fire ant and how I'd decided not to stick my hand in a fire ant mound, "on the theory that a working journalist should strive to be no more than half as dumb as a politician seeking office."

I guess I should have said a *print* journalist because soon after, by some malicious quirk of the fates, I became involved with a television show about fire ants, and almost the first thing I was obliged to do was kneel in front of a camera while sticking my hand in a fire ant mound and saying, "Hello, my name is Richard Conniff." We did about 20 takes, because, after the first few dozen stings, I had a little trouble remembering my name. Also the camerawoman kept talking about how beautifully the light fell on my contorted features. For the next week or so my hand was a mass of welts and pustules.

What I had entered was either the glamorous world of television, or one of the outer rooms of purgatory. And to be here, I had turned down a conflicting assignment to spend three weeks in Provence. The thought still makes me sob. The crew referred to me in the lingo of the trade as "the talent." Usually in a context like, "Why is the talent always such an amazing idiot?" But never mind.

I was under the passing delusion that my future lay in television and particularly in the strange world of natural-history television. (Or maybe, in the standard script language of the genre, I should say the bizarre, mysterious, and/or alien world of natural-history television.) So over the next year my left hand was attacked by killer bees and an Australian box jelly, the latter usually sufficient to cause an agonizing death in less than 15 minutes. Also two fingers got chopped off in transit.

But maybe I need to clarify a few things here. The hand that I stuck in the fire ant mound while attempting to smile for the camera was in fact my own hand. But not even a television journalist is dumb enough to volunteer to be stung by killer bees and box jellies. (Yes, I can think of one or two anchors for whom it might be a close call and many more where it would be a public service, especially if there were earnest interviewers nearby to ask questions like, "Bob, can you tell us what you're feeling at this difficult moment?") Anyway, the killer bee/box jelly victim was actually a prosthetic model of my left arm. It cost $3,000 and it had a slick little pump you could use to make the tendons on the back of the hand arch up in apparent agony.

So I suppose you could say we faked the close-ups.

I am confessing this now because a well-known natural-history filmmaker has been attacked in the press for staging several sequences supposedly shot in the wild. Among other ethical failings, he was accused of "tucking a ptarmigan in his pocket to move it to a more picturesque spot for filming." I do not mean to defend this filmmaker, and I also want to make it clear right now that, in my entire television career, I have never tucked a ptarmigan anyplace.

But the truth is that we filmed most of the fire ant show in the middle of the spare bedroom of an apartment in Tallahassee, Florida.

There were several good reasons for this. From a filmmaker's point of view, the problem with fire ants is that they are smaller than freckles, and when they are not colonizing someone's home, they generally live underground. The only really practical way to film them is on a set. So saying we "staged it" is really a bit of an understatement.

Our cinematography team consisted of a slight, grizzled cowboy in a black Stetson, and his wife, a tall, voluptuous, ginger-haired woman with a certain resemblance to Betty Boop. They holed up in the apartment for several months, sending us occasional progress reports that exhibited a weakness for puns. I have no idea how they explained their "antics" to the landlord. All the windows were blacked out, and the only furniture in the place was a bed. The kitchen door was blocked off by a brace for immobilizing the cowboy's arm during a time-lapse sequence of a pustule erupting. (One imagines the landlord mulling it over with his wife: "Honey, I don't care if they're sado-masochists. They pay the rent.") Colonies of fire ants were thriving in glass-fronted cases, and the air was ripe with the smell of the frogs the ants were eating.

My prosthetic left arm dominated the spare bedroom, with my wedding band slipped onto the ring finger for verisimilitude. Fire ants wandered among the artificial hairs and drove their stingers into the rubberized flesh, which never flinched, except under pneumatic command. The big crisis in filming was the day the exterminator made his rounds, and the cowboy and Betty Boop had to fling themselves bodily across the doorway to keep him out.

Anyway, the film got made, and it premiered on a Sunday night just after a James Bond movie. The Nielsen ratings were respectable, and at least one aspect of my performance caught the eye of other producers: my prosthetic left arm has gone on to mayhem and gore in a series of other natural-history films.

Meanwhile, I am waiting by the phone for another chance at television immortality, and I am keeping my options as a print journalist open.

Family Politics

One afternoon in the 1970s, at a zoo in the Netherlands, a burly, dopey-faced chimpanzee named Nikkie, a veteran of the Holiday on Ice Revue, chased another male named Luit around their compound. Chimpanzee males have five times the upper-body strength of a college football player, and roughly the same sense of decorum. So other chimps leaped for safety as the two combatants kicked up clouds of dust. You could hear their screaming at the far end of the zoo. Nikkie and Luit ended up in the bare branches of a dead oak tree, safely separated, panting as they gradually calmed down.

This raucous confrontation fit the conventional thinking of 30 years ago perfectly. In a lingering echo of World War II violence, many biologists saw higher primates, including humans, as natural-born "killer apes," their lives defined largely by competition, territoriality, and dominance. But the biologist looking on that afternoon at the Arnhem zoo, a soft-spoken young Dutchman named Frans de Waal, noticed something different. About 10 minutes after the fight, Nikkie, the alpha male in the group, reached out his arm toward Luit, fingers extended, palm upward, in an offering of peace. De Waal snapped a photograph and watched as the two apes descended to the fork of the tree, where they kissed and embraced, then groomed each other.

Most researchers would have consigned that moment to oblivion under some dusty academic category like "postconflict interaction." Discussing animals in language that hinted of emotion was heresy then. But de Waal described what had happened between Nikkie and Luit with the same word he would have used after a fight between his own brothers: it was a "reconciliation." Moreover, he went on to argue that chimps are champions of reconciliation. If they could seem at times like killer apes, it was far more accurate to describe them as natural-born peacemakers.

This arresting new view of primate behavior, and de Waal's gift for writing about it gracefully in a series of popular books, have placed him among the leaders of a quiet revolution in our ideas about the animal world and about ourselves. Harvard University biologist E. O. Wilson credits de Waal with "moving the great apes closer to the human level than could have been imagined as recently as two decades ago."

Anthropologist Sarah Blaffer Hrdy, once a skeptic, now calls herself one of de Waal's "biggest fans" and says his description of the tactics primate societies use to stay together "in spite of their dominance-seeking and even murderous tendencies was terribly important, most especially if we want to understand humans." It's a measure of de Waal's influence beyond the scientific community that his book *Chimpanzee Politics*—an account of his years at Arnhem—has been singled out as recommended reading in Congress, and his laboratory in Atlanta has been visited (though with unknown effect) by one of the richest alpha males in the world, Microsoft co-founder Bill Gates.

By watching animal groups in captivity for thousands of hours, de Waal helped show just how narrow and misguided the "killer ape" stereotype really was. His research has also helped correct popular misconceptions about what "Darwinian" behavior means, particularly the "selfish gene" idea that we are born ruthless and, except for human culture, would probably stay that way.

De Waal has argued, on the contrary, that morality is a part of our innate biological heritage. In his 1996 book, *Good Natured*, he didn't quite say that we are born to be good, merely that being good—being cooperative and conciliatory—is the likeliest way to thrive in a social group, whether it's a bridge club, a stock broker-age house, or a chimpanzee troop. According to de Waal, chimps, like humans, live by a highly developed set of social rules. They display a keen sense of fairness in their daily give-and-take, an appetite for punishing individuals who misbehave, empathy for victims of injustice, an interest in peacemaking after conflict and, above all, an abiding concern with the maintenance of good relations in the community. De Waal believes this is the rudimentary basis for morality, which may thus date back more than 5 million years, to the time when chimpanzees and humans shared a common ape ancestor. He feels so positive about the connection that he titled one of his ape books *My Family Album: Thirty Years of Primate Photography*.

Other researchers have since found forms of reconciliation in far more distantly related species, from hyenas to fish. Moreover, when neuroscientists have conducted imaging studies on human subjects, they have found that moral questions light up some of the most primitive emotion centers of the brain. "Morality is not a superficial thing that we added on very late in our evolution," de Waal says. "It relates to very old affectionate and affiliative tendencies that we have as a species, and that we share with all sorts of animals."

Chimpanzee Soap Opera

As in a good soap opera, chimpanzee life is anything but gentle, and, his interest in peacemaking notwithstanding, that suits de Waal just fine. On this lazy spring afternoon, he sits watching his study animals from a boxy yellow tower beside an open-air compound, part of the Yerkes National Primate Research Center at Emory University in Atlanta, where de Waal is a psychology professor. Below, one chimp strolls past another and deals out a slap that would send a football tackle to the emergency room. A second chimp casually sits on a subordinate. Others hurl debris, charge, bluff, and displace one another. One chimp lets out an outraged *waaa!* and others join in till the screaming swirls up into a cacophony, then dies away.

De Waal, going gray at the temples, in round, wire-rimmed glasses and a "Save the Congo" T-shirt, smiles down on the apparent chaos. "Growing up in a family of six boys, I never looked at aggression and conflict as particularly disturbing," he says. "That's maybe a difference I have with people who are always depicting aggression as nasty and negative and bad. I just shrug my shoulders and say, 'Well, it's a little fight. As long as they don't kill each other . . .'" And killing members of their own troop is something chimpanzees rarely do. Their lives are more like one of those marriages where husband and wife are always squabbling, and always making up.

De Waal got his start in biology as a child wandering the polders, or flooded lowlands, in the Netherlands, and bringing home stickleback fish and dragonfly larvae to raise in buckets and jars. His mother indulged this interest, despite her own aversion to seeing animals in captivity. (She was the child of a pet-shop owner and joked that the gene had skipped a generation.) When her fourth son briefly considered studying physics in college, she nudged him toward biology. De Waal wound up studying jackdaws, members of the crow family, which lived around (and sometimes in) his residence. Jackdaws were among the main study animals for de Waal's early hero, Konrad Lorenz, whose book *On Aggression* was one of the most influential biological works of the 1960s.

Like many students then, de Waal wore his hair long and sported a disreputable-looking fur-fringed jacket, with the result that he flunked a crucial oral exam. The chairman of the panel said, "If you don't have a tie on, what can you expect?" De Waal was furious, but during the next six months preparing for his makeup exam, he got his first chance to do behavioral work with chimpanzees. This time he passed the exam, and then threw himself full-time into captive primate studies. Eventually he obtained three different degrees at three different universities in Holland, including a PhD in primatology. He continued his research with chimps at the Arnhem zoo, and then spent 10 years studying macaques as a staff member at a primate center in Madison, Wisconsin. Since 1991, he has divided his time between research at Yerkes and teaching at Emory.

At Yerkes, the chimpanzee compound for de Waal's main study group, FS1, is an area of dirt and grass bigger than a basketball court, enclosed by steel walls and fencing. The chimps lounge around on plastic drums, sections of culvert pipe, and old tires. Dividing walls angle across the open space, giving the chimps a chance to get away from one another. (The walls, says de Waal, "let subordinates copulate without getting caught by the alpha.") Toys include an old telephone book, which the chimpanzees like to shred as a form of amusement.

It is, de Waal acknowledges, a completely artificial environment. Unlike chimps in the wild, his charges don't spend seven hours a day foraging across their home range, they face no competition from outside groups, there are no immigrants or emigrants, and because of a worldwide surplus of captive chimps, birth control is mandatory. Captivity also reduces the power difference between males and females; females who live together defend one another against male aggression. "But the basic psychology of the chimpanzee and the basic behavioral repertoire are still there," he says.

Captive studies also offer one crucial advantage: "You have control and you can see more," says de Waal. In wild studies, it's often a matter of luck whether you find the animals in the first place, "and it's tricky to see when they have a fight, because they tend to run into the underbrush. So to follow what happens after a fight is almost an impossibility." Students of captives used to say that research in the wild was anecdotal and unscientific; the wild researchers in turn said captive work had nothing to do with how animals really live. But the two sides now often collaborate. "I look at it that we need both," says de Waal.

He points out some of the characters in FS1, his mild Dutch accent rendering Rhett as "Rat" and Peony as "Penny." The alpha male in the group is a bristling middleweight named Bjorn. "He's a very hyped-up mean male who has come to the top, I think, by fighting dirty." Within the group, rival males generally fight by something like Queensberry rules. But not Bjorn: "He makes injuries in the belly, in the scrotum, in the throat, places that are potentially dangerous."

These tactics naturally make Bjorn an unpopular alpha. He faces constant pressure from his closest rival, a larger male named Socko, short for Socrates. Nobody else much likes Bjorn either, and they show it by the pant-grunts with which subordinates acknowledge another chimp's superior status. "Bjorn has to work for his pant-grunts," says de Waal. "Socko gets his for free." As in most human groups, regime change is a tantalizing possibility, and de Waal notes, a little hopefully, that Socko's kid brother Klaus is almost at an age when the two siblings might form a coalition to oust Bjorn from power.

Hairy Beasts, Quantified

To the newcomer, all this tends to look like an indecipherable tumult of hairy beasts "chaotically charging around uttering ear-piercing screams," in de Waal's words. It is a little startling to see an observation get recorded, via Palm Pilot, as an equation like this:

> 31 a4 71 a0 w, 036.979
> 82 a4 31 a4 d, 037.461

Translated, this means that Socko (animal 31) trampled or bit (that's an a4, on a scale of antagonistic incidents from zero to six) the subordinate male named Rhett (71), resulting in a win (w), at about 37 minutes into the observation period. Bjorn (82) came charging to the rescue with an attack on Socko, who smacked Bjorn right back with an a4. It's a measure of Bjorn's unpopularity that Rhett responded to this attack by siding with his assailant, Socko, leading to a draw (d). The conflict resolved itself 45 minutes later when Socko began to groom Rhett, with Bjorn promptly joining in. Then Bjorn groomed Socko, with Rhett joining in.

The statistical technique is not that different from taking a football game (more hairy beasts chaotically charging around uttering ear-piercing screams) and logging it into record books as a neat set of numbers. It allows de Waal to sort through the 4,000 or so antagonistic episodes he has recorded over the years to find out how often Bjorn and Socko have fought, how supporting characters shifted loyalties, who won, and whether they reconciled. The statistics turn animal-watching into science, providing a database to test theories.

As with sports, you can always argue about what the statistics mean, and some scientists believe de Waal overstates the case for peacemaking. "I think he is not fully aware of the importance of unreconciled aggressive interactions in the wild—or he chooses not to recognize them," says Richard Wrangham, whose book *Demonic Males* uses ape behavior to explore the origins of human violence. When chimps quarrel in the wild, he says, they reconcile less than 13 percent of the time, only about a third as often as de Waal reports in captivity. This may be because combatants in the wild can get away from each other or because, for strategic reasons, they choose not to reconcile. But Wrangham promptly adds the conciliatory note that he "totally admires" de Waal for having demonstrated that primate life isn't exclusively about waging war, nor about making peace. "If you know chimps, you know they do both."

Indeed, despite the combative nature of academic life, de Waal manages to remain widely liked even among fellow primatologists, possibly because his own behavior is so conciliatory and easygoing. He drives out to the Yerkes field station in an avocado green 1970 Pontiac Catalina, a big boat of a car that he calls "an immigrant type of infatuation." (His other car is a more conventional Toyota.) He speaks English to his students, French at home, German to colleagues passing through, and Dutch to his chimpanzees. He and his wife, Catherine Marin, who was born in the Loire Valley and teaches French at Georgia Tech, live in a wooded area outside Atlanta. They are both workaholics and made a conscious choice not to have children. But they keep goldfish ponds, birdbaths, hummingbird stations, and aquariums around the house. De Waal also bakes bread, and plays Bach and improvised blues on the piano. His upbringing has made him a joker, which he describes as a good survival strategy when you are the fourth son and have no hope of becoming the alpha.

His work has always been distinguished, says *The Naked Ape* author Desmond Morris, by a rare "combination of objectivity and imagination." De Waal knows the names of his chimps, their friends and family, their rivals, their characteristic facial expressions and vocalizations, their quirks of personality. He does not worry about the old rule in Western science against attributing feelings, thoughts, or even individuality to mere animals. Rather, he writes of animals and humans sharing a "vast common ground" of behaviors. "Instead of being tied to how we are unlike any animal, human identity should be built around how we are animals that have taken certain capacities a significant step farther."

Political Primates

Back when de Waal started out, the ban on "anthropomorphism" was intended to discourage naively projecting human states of mind onto animals. But the intended objectivity was often an illusion. Scientists were happy to describe two chimps as "rivals," according to de Waal, but balked at the idea that chimps could also sometimes be friends. (The word "affiliative" seemed to come more readily to their lips.) They projected their own interest in aggression onto their study animals. (In fact, it turns out, chimps and most other primates spend only about 5 percent of their day in aggressive encounters.) Worse, says de Waal, the underlying idea of an absolute divide between the behaviors of animals and humans was paralyzing, a kind of "anthropodenial." It turned the animals into robots, "blind actors in a play" that only we understood.

But in real life, animals don't act that way. At the Arnhem zoo, for example, one of de Waal's favorite chimps was a deposed alpha named Yeroen whose blustering shows of dominance no longer impressed because he needed to sit down afterward "with eyes shut, panting heavily." Yeroen was enough of a schemer to play the younger males off one another and hang on as a king-maker. He allied himself with Nikkie, helping him to become the alpha. In repayment, Nikkie indulged the old fox's sexual forays with females in the group, a privilege the alpha would normally try to preserve for himself. To keep the boss on edge, Yeroen would sometimes side with Nikkie's rival Luit. Yeroen was any-thing but a blind actor in this drama.

De Waal came to view chimpanzee life mainly as an endless round of challenges, fights, coalition-building, and brokering of favors. He had witnessed the treetop reconciliation between Nik-kie and Luit, and he had an idea that peacemaking behaviors were essential to hold the group together through all this maneuver-ing. But the maneuvering still dominated his thinking. "Whole passages of Machiavelli seem to be directly applicable to chim-panzee behavior," he wrote in *Chimpanzee Politics*.

If Machiavelli's *The Prince* had been the first book to frankly describe the power motives and manipulations among the human elite, *Chimpanzee Politics*, published in 1982, was the first to show that these behaviors were embedded in our animal evolution. De Waal's work at Arnhem had taught him, he wrote, "that the roots of politics are older than humanity." It wasn't a case of projecting human patterns onto chimpanzees: "The reverse is nearer the truth; my knowledge and experience of chimpanzee behavior has led me to look at humans in another light."

The idea of chimpanzee politics naturally attracted the inter-est of reporters, who asked questions like "Who do you consider to be the biggest chimpanzee in our present government?" De Waal declined to make such a comparison. "People do it to mock the politicians," he remarked, "but I feel they insult my chimps."

On the other hand, politicians themselves have sometimes seen a resemblance.

When he became speaker of the U.S. House of Representatives in 1995, Newt Gingrich placed *Chimpanzee Politics* on his list of recommended reading for incoming Republicans. Gingrich himself proved adept at fierce infighting, but seems not to have paid as much attention to the parts of the book about reconciliation. He eventually resigned from Congress in the face of subordinate rebellion after the Republicans suffered massive losses in the 1998 elections.

Born to Cooperate?

De Waal, by contrast, went on after *Chimpanzee Politics* to emphasize peacemaking. What struck him as he watched his chimpanzees was that serious injuries rarely occurred during fights within the group. Nikkie and Luit, for instance, never exchanged an actual blow during their brawl before Nikkie's gesture of peace in the oak tree; their only physical contact occurred as part of the reconciliation. Likewise, Bjorn may bite in the wrong places, but he never unleashes the deadly force chimpanzee males can deploy against outsiders. When de Waal was starting out, the best explanation for such restraint came from an eminent evolutionary biologist who concluded "with a lot of fancy mathematics" that animals generally don't try to kill one another because if they did, their rivals might kill them. "It was so simplistic and so limiting in perspective," says de Waal. "It didn't talk about animals liking each other or needing each other or living together. It just talked about fear of injury."

De Waal came to believe that his chimps lived according to a loose system of favors given and received, what evolutionary biologists called reciprocal altruism. The idea that primates may have evolved for this sort of altruism also gave de Waal the means to counter one of the most pervasive concepts in modern biology. In his influential 1976 book, *The Selfish Gene*, Oxford evolutionist Richard Dawkins argued that we are little more than a product of our genes and that these genes have survived by being as ruthlessly competitive as Chicago gangsters. This became one of the most misinterpreted ideas of the 1980s and '90s and, like "survival of the fittest" in the age of the robber barons, helped to rationalize an era of flamboyantly selfish misbehavior. (At Enron, with its culture of gleefully stealing from grandmothers, CEO Jeff Skilling touted *The Selfish Gene* as his favorite book.) In fairness, Dawkins didn't intend it that way. "Let us try to teach generosity and altruism," he wrote, "because we are born selfish."

The animal that first made de Waal see the importance of cooperation was a big, lumbering old female chimp named Mama with an "enquiring and all-comprehending" gaze. She once broke up a fight between two warring males by embracing them, one in each arm. Another time she went up to a screaming male and put her finger in his mouth, a gesture of reassurance. Then she turned to his rival and called him over for a kiss, after which the two combatants embraced. "These males are very tense and they're dominant and strong and aggressive," says de Waal. "So to step in and bring them together is a risky business. To me, it means that she cares about relationships in her community. Chimpanzees have something like 'community concern.' They live in a group and they have to get along, and their life is going to be better if their community is better. That's the selfish motive. But this is also the basis of our moral systems: Our life will be better if our community functions better."

Below de Waal's tower at that moment, Bjorn and Socko are embracing like old pals. It is feeding time, and the Yerkes staff often serves a meal in two or three bundles, to study how factors like friendship or rank affect the tricky business of sharing food among the 20 or so chimps in the compound. So the moment is fraught with excitement and anxiety. Much as two humans declare their good intentions with a handshake—a literal clasping of weapon-hands—the chimps make obvious gestures of nonaggression. Bjorn is allowing Socko to gnaw on the back of his wrist, and Socko is letting Bjorn mock-bite his shoulder. They slap each other on the back affectionately and dance up and down. Then Bjorn, Socko, and Klaus, the three top males, join together in a huddle. It is not quite grace before dinner. But it is not all that different either. The chimps are so intent on defusing the tension of the moment that they completely miss the first round of sugarcane and don't seem to care.

Lest this display of community concern seem a little too rosy, de Waal points out another character in FS1. "Georgia's over here," he says, indicating a 15-year-old female indistinguishable to a newcomer from all the other females. "She's a very mischievous, troublemaking chimp." As a youngster, one of Georgia's favorite tricks when she saw visitors arrive at the FS1 compound was to run to the spigot and collect a mouthful of water. Then she would mingle among the other chimps, lips sealed, doing her best to look like a dumb ape, a blind actor. If the unwary visitors eventually came close enough, Georgia gleefully squirted them, to general shrieking and laughter from the other chimpanzees.

Since evolutionary theorists first suggested it in the nineteenth century, the human connection to animals, and particularly animals like Georgia, has often seemed like an affront to our dignity. For some religious denominations, it was an assault on the doctrine of human supremacy. But when de Waal talks about the primates he has known, it seems reasonable to take comfort in the connection. Beneath the veneer of civilization, we humans are not necessarily killer apes, nor have we evolved to be as ruthless as Al Capone. We are simply social primates, endlessly working out the business of living together. It is difficult business because, like chimpanzees, we are a quarrelsome species. But if de Waal is right, peace may also come to us more naturally than we imagine because of our long evolutionary history of empathy, reconciliation, cooperation, and morality.

Even that reprobate Georgia shows signs of reforming. She is a mother now herself and eager to fit into the give-and-take of the community. She has figured out that the man in the yellow tower wields some influence in her world and greets him lavishly, if only for strategic reasons. She no longer spits at guests. She is still not a perfect lady, just a pretty good chimp figuring out how to get along in her world. Watching them together you start to think that it's not such a stretch for de Waal to call such creatures family.

A Little Sneaky Sex

Somewhere in a rainforest in Panama, a big bruiser of a dung beetle, with a formidable horn on his snout, stands ready to defend his turf. Let's call him Mr. Big. He is, by the standards of his species, the beau ideal: not only tall, dark, and handsome, but also ferocious in combat. What he's defending is, OK, howler monkey flop, but this is an insanely precious commodity for local dung beetles. They get to it 15 seconds after it hits the ground.

At the other end of the tunnel where Mr. Big stands guard, a female is sequestered beneath the monkey dropping. She's supposed to be busily packing up balls of dung and storing them in the larder as food for her offspring by Mr. Big.

Instead, she is having sex with a stranger, a dismal runt named Raoul.

What's wrong with this picture? Absolutely everything, at least according to our conventional notions about sexual behavior. It's part of our lore that the Mr. Bigs of the world—the beefy macho types—get the girls. They also get to kick sand in the faces of the 98-pound Raouls. The stereotype may even seem to make crude Darwinian sense (at least to throwbacks), in that the spoils accrue to the strong.

What Goes on Underground

But recent research has starkly demonstrated that our stereotypes are wrong: The natural world is full of what biologists call "satellite males" or "sneaker males." Many of them are relative weaklings, or lack the masculine ornamentation to dazzle choosy females. Some even practice unconventional strategies like cross-dressing. Yet they manage to defeat the expectations of the macho types: Surprisingly often, it's the 98-pound weakling who gets the girl.

Research over the past two decades has documented sneaker male behavior as a standard tactic in hundreds of species, from damselflies to white rhinos. Among red deer, for instance, a 12-point buck may be the apparent stud, defending a harem of 20 hinds. But groups of young males with dinky little antlers loiter nearby and grab occasional matings. (British biologists refer to them as "sneaky fuckers.") Similarly, an old horseshoe crab may no longer have what it takes to latch onto a female of his own. Instead, he waits to find a young couple already mating, then scrambles aboard, spills his seed (well, *gross*, to be sure), and manages to fertilize about 40 percent of the female's eggs.

But let's go back for a moment to Raoul, whose sneaky behavior is considerably more appealing: Mr. Big is twice Raoul's size, with a horn that accounts for up to 15 percent of his body weight. Among dung beetles, the horn is the main male secondary sex characteristic and the chief weapon for head-butting combat. Raoul, by contrast, has no more than a pitiful nub where his horn's supposed to be. Having been malnourished at a crucial stage in his early development, he will never grow a respectable horn, nor ever be able to stand in the doorway of his own handsome pile defending his claim to a female. About half of all male dung beetles suffer the same sorry fate. To the old way of thinking, they could just about resign themselves at birth to romantic and evolutionary oblivion.

"But nobody had looked at what goes on underground," says Doug Emlen, a biologist at the University of Montana. He set out to change that as a graduate student at the Smithsonian Tropical Research Institute in Panama. Emlen built glass-fronted ant farms, put the dung beetles inside, and then watched what really happened. The smallest sneaker males, he says, relentlessly attempted to slip into the tunnel entrance while Mr. Big was looking the other way. Other sneaker males, like Raoul, withdrew to a respectful distance and dug tunnels of their own. Once hidden safely underground, these sneakers veered sharply sideways to intersect the main tunnel and enjoy a tryst with Mrs. Big while Mr. Big himself stood stupidly at the door.

Emlen's supervisors were dubious: "Maybe that's just something that happens when you put dung beetles in ant farms?" they said. So Emlen went out and made latex casts of dung beetle tunnels in the wild. What he found was that, after Raoul kisses Mrs. Big good-bye, he often continues digging sideways to intersect the tunnels of five or six other females, in effect trap-lining for love. Not only do these sneaker males get the girl, but they may also be more likely to fertilize her eggs because, in at least one dung beetle species, smaller, sneakier males have larger testes and produce more sperm.

It was enough to make the Vin Diesels of our world whimper in their sleep and reach out for their wives.

The Joy of Cross-Dressing

Up until the 1970s, it was scientific dogma that the males in every species had one ideal life history, one form most likely to charm female counterparts. Little boy dung beetles and damselflies all supposedly wanted to grow up to be fighters. At least in theory, the combined forces of natural selection and sexual selection would ruthlessly weed out the sissy alternatives.

Darwin himself noticed that alternative forms in fact persist in almost every species. But biologists generally overlooked this phenomenon because they couldn't explain it—or because they just didn't see it in the first place. Sneaky behaviors designed to fool other members of the same species often fooled outside observers, too: Biologists tended to notice mainly conventional males, and even then mainly *successful* conventional males with territories and harems. The scientific evidence was thus skewed from the start to support the fighter-male hypothesis. If alternative types got occasional matings, it was only because they'd gotten lucky while making "the best of a bad job."

"Theory determines what you see," says University of Toronto biologist Mart Gross. And in the political and scientific zeitgeist of the late 1970s, people started to look beyond the fighter stereotype. Evolutionary theorists were just coming to terms with the idea that, for two or three different tactics to persist in a significant portion of a species, they ought to be equally good. Otherwise, evolutionary selection would drive out the inferior ones. But it required a kind of collective gulp to face up to the question of how a male might actually benefit from being smaller, weaker, and less conventionally masculine, if not downright effeminate.

For instance, in bluegill sunfish—commonly known as "sunnies" to any kid who's ever gone freshwater fishing in North America—a large percentage of males impersonate females. They wear female striping, and their eyes dilate to resemble the limpid black pools of a girl bluegill. They also have heightened levels of the female hormone estradiol. Gross coined the term "satellite males" to describe the way these males orbit the nests of conventional males.

The conventional males frequently get fooled and invite the cross-dressers into their nests. Gross carefully documented what happens next: When real females visit a nest, they dip down, turning on one side to release eggs for the resident male to inseminate. When a female-impersonator enters a nest, he dips in exactly the same way. But—*vive la différence*—the cross-dresser is in fact secretly releasing his sperm onto eggs left by the real females.

This kind of transvestite mating tactic has turned out to be surprisingly common in other species. For instance, the giant cuttlefish is a voluptuous relative of the squid, 3 feet in length. Conventional males in this species are the John Travoltas of the animal world and put on a lavish, romantic display for females. A courting male waves the long, bannerlike webs on his arms and turns his body into an amorous light show, with bright colors and pulsing zebra stripes. A smitten female will swim alongside, pulsing back, the two of them performing a kind of neon-lit courtship dance. A small sneaker male will also sometimes escort the happy couple, and by assuming the shape and body patterns of a female, he avoids getting attacked by the courting male.

The sneaker male's strategy is simple: Sooner or later, Courting John will get distracted and have to drive off some other big male, who seems like the more obvious rival. In the resulting confusion, says Australian biologist Mark Norman, the female-impersonator "waltzes in, sneaks under the covers, grabs the female, and starts passing sperm to her." When Courting John comes back, the sneaker male once again tags along looking girlish and innocent.

Cross-dressing in the animal world isn't just a way to get a mate. Some female damselflies appear to imitate males for the exact opposite reason: To avoid being relentlessly courted and sexually harassed. And some male garter snakes impersonate females during the mass springtime emergence in Manitoba, Canada, apparently because . . . well, it's damned cold in Canada at that time of year, eh?—and just a little less cold underneath a swarm of eager (but deluded) conventional males. That's their story, anyway. Besides, the bottom of a swarm is a good place to hide from predatory crows. It's also possible that some animals, like some humans, imitate members of the opposite sex for the sheer thrill of it, but we have no evidence for that.

The Milkman's Children

Relentless interloping by sneaker males, combined with direct *mano a mano* challenges by conventional males, mean that the big boys of the world often spend an inordinate amount of time and energy on what biologists call "mate guarding." (Also known to human females as "desperate male clinging.") Among dung beetles, for instance, Mr. Big doesn't merely stand guard in the doorway, but also regularly patrols down the tunnel to check on Mrs. Big, and if he catches her *in flagrante delicto*, he throws the interloper out. Then he immediately has sex with Mrs. Big, to displace the interloper's sperm with his own. In fact, he may have sex with her 10 times a day, just to be certain—or at least a little less uncertain—that he's the father of her offspring. (Waterbugs are even more anxious about paternal uncertainty, because they literally have to carry the female's offspring on their backs. In one study, a male waterbug mated with the female 100 times in 36 hours.)

The burden of mate guarding, whether by fighting or by marathon love-making, can leave even the sturdiest male feeling weak-kneed. Among northern elephant seals, a big bull that wins bloody contests against other bulls gets to stake out a stretch of beach and wallow in lubricious splendor with his harem. He may mate with 50 females in a season, while many lesser males go celibate. But defending his harem also means not going back to the sea to eat, and he typically loses about 40 percent of his weight, roughly equivalent to slimming down from a Ford Explorer to a Volkswagen Beetle, during the three-month breeding season.

This is the quintessential male dilemma: how do I fill my belly and not come home to a nestful of the milkman's children? And it suggests why success as a conventional male may not always be all it's cracked up to be. DNA testing in red-winged blackbirds, for instance, has revealed that the longer a resident male stays away from the nest gathering food, the greater the likelihood that his putative offspring will actually be fathered by someone else. In one Australian species, the superb fairy wren, fully two-thirds of the offspring get fathered by somebody other than the man of the house. So why bother to have a nest in the first place?

If male animals could perform a cost-benefit analysis, they would almost certainly conclude that it's cheaper to be a sneaker male: You don't need to keep up a large territory, court females, fight off rivals, or provide parental care. The possible charms of the footloose lifestyle loom large for every male, particularly around the time the mortgage payment comes due. And yet it's hard to believe that sneaking along the outskirts of more conventional society is ever really better.

In most species, females are the choosers when it comes to mating, and they seem to prefer conventional males. Northern elephant seal females, for instance, clearly regard the big beefy alpha as their best possible mate. They howl in protest when the alpha mounts them—but they howl louder and longer if a lesser male tries it, so the big bull will come thundering to the rescue.

From a Darwinian perspective, females exercise sexual selection to obtain the mate with the greatest possible fitness—and in elephant seals and other species, he often looks a lot like Mr. Big.

Season of the Slacker Male

So when is cheap-and-sneaky actually better? Romance is a strange business, and changing seasons and circumstances can make it stranger still. Side-blotched lizards, for instance, live among rocky outcrops in the coastal mountains of central California. They're about 2½ inches long, and conventional males typically display by doing pushups while also puffing out the bright patch of color at their throats. Blasé females watch skeptically and murmur, "Oh, yeah." Barry Sinervo, a biologist at the University of California at Santa Cruz, has identified three different genetically determined mating tactics, each conveniently identified by a distinctive throat color. It's a bit like basketball teams in different-colored jerseys.

The Big Boy Orange-throats are the "ultradominants," patrolling large territories with lots of females. The Average Joe Blue-throats are about 15 percent smaller, but still manage to maintain modest territories and diligently mate-guard however many females they can round up. And then there are the Slacker Yellow-throats, which mimic females, do no pushups, and maintain no territories, but instead skulk in nooks and crannies among the rocks hoping to get lucky. It ought to be easy for everyone to spot the Slackers and make them disappear.

But is life ever that simple? If Average Joe Blue-throats happen to predominate in a population, then the Slacker Yellows are indeed out of luck. The Blue-throats recognize them as rival males and boot them out of their territories. But Blue-throats never predominate for long, because the Big Boy Orange-throats have the size and stamina to beat them up and take away their females. Thus Orange-throats get more offspring in the next generation, and in a year or two, they invariably supplant the Blues.

But Orange-throats are apparently not so bright: They don't recognize the Slacker Yellow boys as rival males, so they tolerate them in the neighborhood. This allows the Yellows to make what Sinervo calls "little sperm strafing runs" through Orange-throat territories. After a year or two, Yellow slackers actually outnumber Orange-throats. Then it's time once again for the Blue-throats to make their move. In fact, the three male types regularly displace one another in the reproductive hierarchy over a five- or six-year cycle. Changing circumstances mean that the mating tactic that's lord of the dance one season is out on the street the next.

In many other species, the sneaky approach seems to work best when picky females concentrate their favors on a relative handful of big brutish males (or when the big brutes corral females in harems). The alpha males do best with this kind of strong sexual selection, and the overwhelming majority of conventional males wind up as losers. But sneaker males—the cross-dressing harem-raiders of the world—may actually do better on average than the ordinary Joes. "Any time you have a great imbalance between the haves and the have-nots," says Purdue University biologist Richard Howard, "the have-nots find a way around." Sneaking also works better in habitats with lots of hiding places to run yipping and screaming for cover when the big guy comes *fe-fi-fo*-ing onto the scene.

At the very least, sneaky mating tactics provide a way for weaker males to avoid being weeded out. But they may actually do much more than that. After his work on bluegill sunfish, biologist Mart Gross went on to study Coho salmon. Some Coho males mature early, at two years of age, and become "jacks," typically measuring less than 15 inches in length. Others delay maturity for an extra year and become big "hooknose" males 25 inches or more in length. Sports fishermen naturally regard the hooknoses as the trophy fish, the best of the bunch. But Gross says they are not.

Among the young males, it's the larger, more precocious ones that typically opt to become dinky little jacks, and Gross argues that "they would not be sacrificing body size unless there were significant advantages" in the sneaker male lifestyle. Both jacks and hooknoses mature in the open ocean. But jacks spend much less time at sea. So the percentage of jacks surviving and making it home to the spawning grounds is almost double that of hooknoses. And once they get there, the jacks are adept at hiding and darting out after a spawning female to steal fertilizations from the lumbering hooknoses. "What we have always assumed to be the best male in a population could be wrong," says Gross. "In a number of cases, it's the alternative." That is, the sneaker male may be the real trophy.

So what does all this mean for humans? To risk scientific and pop-cultural heresy for a moment, is it possible that the real love god was not, after all, the deep-voiced "walrus of love" Barry White ("I'm Qualified to Satisfy You") but some sneaker male like the 5-foot-2-inch pop artist Prince? Not Arnold Schwarzenegger but Danny DeVito? Not Gary Cooper but skinny little Frank Sinatra? You can imagine the headlines: "Ninety-Eight-Pound Weakling Wins Beach Beauty" or "Junior Doesn't Want to Grow Up Big and Strong, After All."

Alas, scientists do not usually go there. They are properly cautious about extrapolating from other animals to explain our behaviors. Scientists might look at riding instructor James Hewitt having an affair with Princess Diana, or actor David Spade bouncing from Heather Locklear to a Playboy Playmate half his age, and mutter, "sneaker male." But the most they will say out loud is that human mating strategies are at least as diverse as those being studied in the rest of the animal world.

In any case, our old stereotypes are unlikely to vanish overnight. And here is the thing: The Raouls and Slacker Yellowthroats of the world probably like it better that way. The tendency of Big Boys everywhere to regard them as harmless is their best possible protection. So relax, big guy. Put your feet up, Shaquille. Settle in with your stereotypes, Clint Eastwood. You should know that the little guys of the world wish you nothing but sweet dreams.

Swamp Thing

It was a fine, gone-fishing sort of morning in May, and we were out on a lake in northern Louisiana peeling leeches off snapping turtles. The tupelo gums and bald cypresses grew close together, and the water around the trunks mirrored each tree perfectly, so our boat seemed to be suspended in middle space, a forest underfoot as well as overhead. The sun slanted down into the shadows.

J. Brent Harrel was trying to reason with a snapping turtle caught in his hoop trap: "I need those leeches, big girl, but you're not going to give 'em to me, are you?" The turtle hissed back. Harrel reached in to extricate her from the net, and she lashed out with her hooked jaws, her pink mouth gaping and her forefeet lurching clear off the bottom of the boat. The jaws clapped shut like a bear trap.

Harrel, a wildlife biologist, deftly flipped the turtle onto her back. He planted the fingers of one hand in the soft yellow flesh under the chin to hold her down while he harvested a glistening bouquet of leeches at her throat. The leeches were a favor for a research colleague, to keep things from getting dull. But Harrel's real passion was for snapping turtles.

Harrel held the turtle up for inspection. Her mossy brown shell was about a foot long and deeply serrated on the back edge. Her wattled skin thickened into a saw-toothed ridge on the tail and leather plating on the legs. She was built like a tank, with a rocket-propelled neck in place of a cannon. She watched for an unwary limb to drift into range. Her sharp, curved claws reached back to scrabble at Harrel's wrists. "You want to pay attention to what she's doing all the time," he advised, "'cause if you forget, she'll remind you."

Big Girl was actually the smaller of the two snapping turtle species prowling around at that moment in the forest beneath our boat. She was a common snapper, *Chelydra serpentina*, which means "snakelike swamp thing." Though you may never see them, even when you swim with them, common snappers are everywhere in streams, lakes, and little ornamental ponds from Colorado eastward and from Ecuador to Nova Scotia. (Some scientists argue that the geographically isolated Central and South American common snappers are a separate species.) The adults typically weigh 10 pounds or more. But they can live for decades, and the largest wild one on record weighed more than 76 pounds.

The other, larger species for which we had set our traps was *Macroclemys temminckii*, which means "Temminck's big turtle." It's better known as the alligator snapper, or loggerhead, and large males can weigh more than 200 pounds. It was once so popular as an ingredient for soup and turtle sauce picante that it is now imperiled throughout its range, which extends from Texas to Florida and up into Illinois.

Snappers are a breed of loners. In their own world they live quiet as rocks, eating everything in reach, from crayfish to skunk cabbage. Alligator snappers actually lure their prey, wriggling a wormy pink appendage in the floor of their gaping mouths as a come-on to visitors. Common snappers appear to rely mainly on their natural camouflage to bring unsuspecting prey within range.

I peered, not too closely, into Big Girl's hostile little eyes, and the word "atrabilious" came to mind. It means gloomy and ill-natured. Just then Harrel noticed that trap number two was rocking violently in the tea-black water, and we put Big Girl in a burlap bag and paddled on.

American Original

I'd learned the word atrabilious only a few days earlier up in New Hampshire from a friend named John Rogers, who travels the Northeast every summer catching common snappers for a living and listening to poetry tapes for self-improvement. It sounded to me like a good word for these malevolent-looking creatures, but I was making the usual mistake with snapping turtles. People are sometimes atrabilious, Rogers said, but gloominess and malevolence aren't snapping turtle traits. Snapping turtles are merely quarrelsome. They are—he hit on the right word—fractious. You would be, too, if someone had just plucked you out of your natural world.

Rogers, who is sometimes known as the king of the snappers, could speak with a certain authority on the fine points of snapping turtle character. By his own estimate he has caught more than 100,000 common snapping turtles over the past 40 years to supply the meat and soup trade. He has also reseeded his hunting grounds over the years by releasing 250,000 small ones into the wild. He has the scars of eight snapper bites on his right hand alone.

A tall man, with a grizzled beard and a greasy pith helmet, Rogers is one of those convoluted people who deeply love the animals they hunt, to the point that predator and prey begin to resemble one another. It was early May, the start of mating season. "You'll see turtles moving all day long now," Rogers said, as he studied a pond with binoculars. "What they are is horny males, full of testosterone, looking for a female or a fight." He grinned. "Just like guys hitting bars on Friday night."

For Rogers and other admirers (many of whom are in prison, to judge from the letters I received the last time I wrote on this topic), the likable thing about snapping turtles is that they are, in a word, trouble. You would not put one in a petting zoo unless maybe you wanted the children to learn to count without using their fingers. On the other hand (a phrase one uses advisedly), they are interesting trouble—tough, reclusive, and fiercely independent, unhuggable in a culture determined to make all animals cute, paragons of the "Don't tread on me" spirit in a society that thinks nature ought to be approachable. Snapping turtles are throwbacks not merely to the dinosaurian epoch during which they evolved but also to our own past as a nation. They are hardheaded American originals.

Negative ideas about snapping turtles nonetheless abound. Teddy Roosevelt once called them "fearsome brutes of the slime," and a large snapper was "the demon of the deep" in one popular children's story early in the twentieth century. Even people who consider themselves environmentalists sometimes kill them because they think snappers are taking all the game fish or because they have seen a line of adorable ducklings get dragged down one by one in a tumult of downy feathers. But like any predator, turtles mainly pick off the weak, helping to strengthen the population of survivors. They also scavenge carrion. "If you were a vegan and you wanted to bribe me," Rogers offered, "I could take you to places where the snappers are all vegetarians."

Other places, of course, they eat almost nothing but flesh, and this may be the real source of human hostility toward them: When people see a big snapping turtle come clambering up out of their favorite swimming hole, they suffer a wildly misguided fear for human life and limb. Not long ago the residents of a wealthy neighborhood in my hometown showed up at a town meeting determined to shut down a public river access; they said visitors were liable to be attacked by monster snapping turtles lurking in the shallows.

But when they are in their own element, snapping turtles almost never bite people. They stick to whirligig beetles and other mouth-sized prey. After Woodstock, the 1969 music festival, Rogers visited the pond where all those free spirits went skinny-dipping and collected dozens of snapping turtles, averaging 20 pounds apiece. Analysis of the stomach contents produced no human body parts. Snappers bite because they cannot retreat entirely within their shells. But even on land they bite only in self-defense and only when someone is dumb enough to fool with them.

Being dumb enough myself, I tested the old lore that a snapping turtle can bite through an oar. My experimental subject, a large alligator snapper, merely scarred the wood. Common snapping turtles typically cannot even bite through a pencil. The hooked jaws of either species will, of course, cut through flesh like tub margarine but generally stop when they hit bone.

When the subject of snapping turtles comes up, the conversation nonetheless turns inevitably, almost ritually, to tales of maiming and dismemberment. "My neighbor had a dog, about a 90-pound Labrador," a Louisiana ostrich farmer told me, "and one day we found him down by the water with a big chunk out of his backside, and it was the exact shape of a loggerhead turtle's jaw. We figured he was dead. But they took him to the vet and got him patched up, and he walked around after that with half a rump." He still ventured into the water, the farmer said. But if so much as a tupelo gum fruit hit the surface, that dog lit out for high country.

In turn, I told the farmer about a Connecticut man who cut off a snapper's head and buried it in a field. A day or two later the man's dog dug up the head, and it bit him on the nose.

Either story might be true: A researcher recently identified the bones of a full-grown raccoon in the scat of a 102-pound alligator snapper. And snappers can indeed still bite after they have been killed. In Louisiana I watched a fish-market counterman take the head off a snapping turtle with a band saw. Then he tapped the turtle's snout with his knife, and the decapitated head clamped shut reflexively on the blade. He held it up for display and said, "This thing'll bite tomorrow."

Snapping Turtle Hillbillies

The tradition of eating snapper meat is strongest in Cajun areas of Louisiana. The fish-market man told me that he sells snappers mainly on Wednesdays and Fridays because of an old custom among Catholics that snapper was an acceptable substitute for meat on days of abstinence. "It don't taste like chicken," another turtle dealer told me, "and it doesn't taste like pork. It's that little thing in between." Otherwise the market for snapper meat is quirky and small, extending up the Mississippi into Illinois and Ohio, out to the snapper soup makers of Philadelphia, and among Asians in New York City and California, who regard turtles as an edible symbol of long life.

Supplying these markets is a small-scale, backwoods sort of business. "The trappers sometimes remind me of the Clampetts," a woman named Sheila Millard Perry told me, the Clampetts being television hillbillies. Perry presides over Millard's Turtle Farm of Birmingham, Iowa, the largest snapping turtle dealer in the nation. Then she added: "My family sometimes reminds me of the Clampetts." The business operates out of a ramshackle assortment of sheds, trailers, and rough-dug ponds in the rolling Iowa hills, and Perry, who used to be a nurse, was wearing denim short shorts rolled up, a torn T-shirt, and white rubber galoshes. The family skinned 137,158 pounds of snapping turtles in the past year, she said. At the time, she was buying live turtles at 60 cents a pound and selling at $3.50 on the bone, $5.75 boneless. It did not appear to be a get-rich-quick sort of business.

Perry took me on a tour of the property to show me some of the turtle by-products. Big shells were drying in the sun on a shed roof, to be sold to arts and crafts types who make them into clocks, saddlebags, breast shields, and Indian dream catchers. The smaller ones become rattles and drinking flasks. The belly plates get turned into knife sheaths. The liver becomes catfish bait. Paws get sewn into pouches. Claws become necklace parts, and a couple of throat bones get glued onto muskrat skulls to sell as Texas longhorn car-mirror ornaments at $3 apiece.

"My father says the only thing out of the snapping turtle that he hasn't figured out how to sell," said Perry, "is the snap."

Researchers have argued that snapping turtles cannot sustain a commercial trade because they are slow to reach reproductive age, and when they do, they lose most of their nests to predators. The females usually bury their eggs in a dirt bank in May or June, and raccoons, skunks, and foxes promptly dig them up. A raccoon will sometimes treat a snapping turtle on the nest like a vending machine, sitting behind her and palming the eggs up into its mouth as they come out. Even if they survive to emerge in August, the hatchlings, which are the size of milk-bottle caps, may get eaten by birds, bass, and other predators.

Even so, common snapping turtles appear to be thriving in most of their range. This may be because the market for snappers is small, and there is no incentive to overharvest. Rogers says he typically leaves a pond idle for five or six years between harvests to allow for recovery. But he argues that the turtles also do well on their own. Over more than 3o years, Rogers told me, he has taken 36,ooo pounds of common snappers out of a single 2oo-acre waterfowl refuge in Massachusetts. A week after we talked, he went back to the refuge and found another thousand pounds of turtle ready for harvest.

Wrestling an Angry Fat Woman

Down in Louisiana, where the alligator snappers are not thriving, Brent Harrel was making a careful study of trap number two. "This is going to be an ordeal," he said, indicating the large male alligator snapper inside. "These boys are tough to fool with. When we pull him up, he's going to be real aggravated."

It took both of us hauling on the hoop net to bring him to the surface, because of his weight and because he was lurching and tearing at the net. We drew him up a little more, and he suddenly clamped his jaw on the gunwale as if to shred the aluminum. We hesitated, regarding each other with raised eyebrows, and then hauled the net over the side into the bottom of the boat. The tattered remains of the baitfish swung into range, and the turtle lunged again. His mouth closed on the skull of a buffalofish. The hollow sound of bone caving in echoed around the boat.

"This is real stressful for him," Harrel said, which, under the circumstances, was either wonderfully magnanimous or pure projection. Harrel reached in and grabbed hold of a hind leg. Drawing the turtle back out through the narrow throat of the trap was like trying to wrestle an angry fat woman out of an undersized girdle. But he succeeded after a while.

The turtle's shell was 2 feet long, and he weighed 95 pounds. He had a smooth, wizened snout and a head like a rottweiler, 23 inches around. We flipped him onto his back, and his wrinkly, tubercled underflesh was stained rust-colored. The plastron was smooth as an old stairway with long use. He made a low, irritated hissing sound, like a scuba diver exhaling.

"Can you just imagine how stressful this is for him?" Harrel said again. "You realize that turtle's probably never been out of the water in a century?" Alligator snappers are thought to live at least that long, and we speculated about whether McKinley or Roosevelt might have been president when this turtle hatched. "The thing I hate when somebody kills one to eat is how old it is," he said. "To harvest a hundred-year-old animal for 40 pounds of meat doesn't make sense."

The rest of the traps yielded two platter-sized softshell turtles, five more alligator snappers, and a mess of red-eared sliders. "Just think what a trapper could do here," he said. Louisiana's turtle-trapping regulations are lenient, and the alligator snappers mostly exceeded the minimum 15-inch shell length. "It'd be a pretty good piece of change for an ol' boy who may not have a high school education. When they find 'em like that, they don't let up, they just keep catching 'em till they exhaust the supply."

We piled most of the catch into a pickup truck to be hauled into town, weighed, measured, and tagged, then returned to the lake later in the evening. But we didn't take the big alligator snapper. Harrel thought the biology department at Northeast Louisiana University, where he earned his degree, might want to keep him to observe in one of its holding tanks. He was also a little afraid that he might want to keep the turtle himself, just to study for a day or two. He wrestled with his Louisiana heritage: "My mama probably wouldn't let me back in the house if she knew I had 'em and let 'em go," he said. He debated at least tagging the turtle for future recapture. But there was an element of desecration in this idea: "He's great the way he is."

So we put the turtle back into the boat and paddled out among the gum trees. We eased him over the gunwale, and the moss began to float up again on his head. Harrel held onto the rear edge of the shell and the turtle kicked water back on us, and then calmed down. "All right, this is it, big boy," he said. A thread of reluctance in his voice, he added: "He'll be gone forever."

Then he let go. The turtle headed straight for the bottom and vanished. A moment later he reappeared, walking across a high spot, a huge, improbable creature, yellow-brown under the tea-colored water, the sun shining on his head. I thought of something John Rogers had told me: "They're beautiful when they're moving in the water. Graceful. Powerful. Majestic. Walking along the bottom, not afraid of anything. They're the king of everything down there. They've been there forever. They were watching dinosaur droppings fall in the water and fertilize the world."

The turtle vanished again into the depths, and we followed his progress for a while by the air bubbles and prickly gum balls his footsteps sent up from the bottom. He passed between two trees and then out into the lake, leaving behind only a long trail of effervescence.

Backyard Wildlife

> *It takes a man of genius to travel in his own*
> *country, in his native village; to make any progress*
> *between his door and his gate.*
>
> —H. D. THOREAU

The scientists launching themselves into Hartford, Connecticut's 695-acre Keney Park at 3 p.m. that Friday were armed with spotting scopes, sweep nets, pit traps, scalpels, and fish stunners. They were prepared to dance like butterflies, sing like chickadees, or do almost anything else this scavenger hunt required. One collector was seeking someone to donate bait for his dung beetle trap. Another was erecting what looked like a device for extracting sunbeams from cucumbers.

It consisted of a big, galvanized-steel funnel mounted on a tripod, and it was actually meant to extract animals from soil samples. With a coffee can, a slight, white-bearded man named Carl Rettenmeyer cut out a disk of grassy soil and leaf litter "about the size of a Quarter Pounder, with lettuce." This sample would sit overnight on a screen in mid-funnel, with an electric light overhead to make things uncomfortable, and a killing jar below to collect whatever came creeping out. By Saturday afternoon, the "Quarter Pounder" of lawn alone would yield 23 separate species, and 89 individual animals, including mites, thrips, and an awful lot of springtails, "little bitty things smaller than the head of a pin." The springtails run around with their tails tucked under their abdomens, spring-loaded so that, in case of danger, they can escape with a backflip, 6 inches into the air. It is a skill two-legged city-dwellers can only envy.

"The area of the sample is 11.044639 square inches," Rettenmeyer said. "Now multiply that by . . ." Then he came up with his best estimate: that there are 50 million individuals in just the top inch of soil in an ordinary acre here, and 35 billion animals in Keney Park—nearly six for every man, woman, and child on Earth.

"People are hearing that word 'biodiversity,' but they think it's in the rain forest," said Ellen Censky, who was organizing the event as director of the Connecticut State Museum of Natural History at the University of Connecticut. "The BioBlitz is a way of letting them know there's a lot of biodiversity right here."

A BioBlitz is an event in which dozens of scientists fan out across some unlikely habitat, hell-bent on recording every species they can find, dead or alive, in a 24-hour period. It's also a lark, the closest most scientists will ever get to a varsity sport. Censky is a wry, enterprising sort who once proposed a museum exhibit that consisted of a dead elephant in a room full of flesh-eating dermestid beetles. It didn't happen (some nitpicky problem with ventilation), but it showed a certain flair for ecological entertainment.

BioBlitzes have become regular events around the country since the first one took place in 1996 in Washington, DC. Censky's team was hoping to top the record of 1,905 species found in a previous biodiversity survey around Walden Pond, outside Boston. She had assembled her roster with a coach's eye for covering all positions, from fish-squeezers to bat-grabbers. She was piling on entomologists with a knack for quickly identifying obscure insects. She had specialty teams, too, including an algae expert and a parasitologist.

Out in a wooded section of the park, a herpetologist named Hank Gruner had just caught a garter snake, which promptly regurgitated a wood frog. Two species for the price of one. Gruner started flipping over old furniture someone had dumped. "Herpetologists love debris," he said, snatching up a writhing, glossy-purplish creature in his palm, his twenty-fifth red-backed salamander in the past hour. "By biomass, they actually outweigh the birds, and they're equal to the small mammals," he said, letting it go again. "All that with no lungs. They breathe by diffusion through their skin and through their throat."

Elsewhere in this depressed neighborhood in Hartford's north end, the algae expert found a green alga closely related to the one from which all land plants on Earth appear to have evolved. *Coleochaete* grows on cattails and rushes, she said, and it's probably 500 million years old. Keney Park was also home to freshwater sponges, as well as to symbiotic algae, which live on the sponges and spoon-feed them the by-products of photosynthesis. The larvae of two species of spongilla fly feed, in turn, on the sponges, sucking up a soup of algae and cellular fluids. One of the fly species moves away from the sponge just about the time of year the other arrives to make its home there, like vacationers time-sharing a condo.

At 10 p.m. at a shallow stream spanned by a 30-foot-long mist net, a bat in the thrall of the hunt got its leaflike wings tangled in the netting. A biologist delicately extricated it, identified it as a big brown bat (*Eptesicus fuscus*), and tucked it into a cardboard toilet-paper roll. The roostlike snugness calmed the bat down long enough for it to weigh in at 13 grams before fluttering free.

Elsewhere in the woods, a black light cast a faint purple glow among the dark tree trunks. On a reflective bedsheet nearby, a

host of light-dazzled wolf spiders, water bugs, beetles, and slug
moths had assembled. The entomologists set to work, and the
air was filled with the sound of killing jars opening and the smell
of ethyl alcohol. A beetle expert got on his knees and sucked up
specimens, *pttt-pttt*, into an aspirator jar. Someone told a story
about a scientist who didn't realize he was working with a defective aspirator. "He wound up with 70 insects in his sinus cavity. An entire ecosystem. Alive. They published an article about
it." (In fact, the sinus had served as an incubator for insect eggs
that passed through the aspirator's fine mesh screen. And having matured there, three adult rove beetles, 13 fungus gnat larvae,
three parasitic wasps, and about 50 springtails crawled to freedom by way of their host's nostrils. It was a clear case of getting
too close to your work.) The beetle guy went *pttt-pttt*, undaunted.

Things quieted down till just before dawn. Then the birders
came out and stood around, at odd angles to one another, heads
cocked, hands in pockets, listening. "It sounds like a worm-eating
warbler," said Frank Gallo, who clearly wasn't expecting to find the
species here. "Before I write this down I want to hear it better."
He bashed in through the undergrowth and called out *pish-pish-
pish*, engaging the bird in a short dialogue. "Pine warbler," he pronounced, satisfied now. Next Gallo did an admirable imitation of a
chickadee, bursting out in a high nasal *deee-deee-deee*. Other species flock around when they hear a chickadee call, he explained.

Good sightings began to pile up as the morning grew long:
a bald eagle, a 12-inch-long pileated woodpecker, a coyote. At
a table back at headquarters, a parasitologist picking through
the guts of a grasshopper came up with two parasites practicing
syzygy, which means that they were mating head to tail. One scientist triumphantly picked a tick off someone's leg, and when a
cat-mauled short-tailed shrew showed up just outside the door,
the maggots got counted, too.

As the witching hour drew near, lepidopterist Dave Wagner had 295 species of moths and butterflies. "I wish I could get five more," he lamented. "How much time do I have? Twenty minutes?" He started to pick through the insect refuse piles where others had been working. "Twenty-seven ants!" someone yelled.

"No more insects," said the woman who was keeping the insect tally, 60 seconds from the 3 p.m. deadline. "No more nothin'."

"I'm up to 305," said Wagner.

"That's it, Dave. We're not counting any more."

". . . 306 . . . 307."

"Wait! I haven't given you my dung beetles yet," another entomologist yelled, fondly cradling his entire collection in a jar lid.

A small crowd had gathered outside to hear the final count for this BioBlitz. It came to 1,369 species. Less than the record at Walden Pond. "Wait'll next year," someone muttered. (Since then, the BioBlitz record has climbed upward of 2,500 species.) "I know some of you won't be happy to hear this," the parasitologist was telling the crowd, "but we were delighted to find 35 species of parasites. So congratulations, Hartford." And from the enthusiasm in her voice, you could tell that the weird little creatures in our own backyards, even down to the lowliest flatworm, were as glorious to her as a pride of lions.

On the Track of the Cat

*T*he first time I got a good look at a leopard in the wild was on a nighttime game drive in Botswana. She was lying in the grass when the spotlight picked out her profile. Her massive shoulder muscles bulked out like a Janus face at the back of her neck. She got up and walked toward our vehicle, where we sat in the open, and looked straight up at us as she slinked past, bathed bloodred by the filter on our light.

The guide recognized her as the daughter of a female he routinely followed. She was two years old, and the big, lumbering Land Rovers of the nearby Chitabe Camp were as natural to her as elephants. She began to hunt, with three vehicles idling 50 yards behind her, through a thin forest of mopane trees. When the slink shifted unmistakably to a stalk, all the engines shut down and the lights went out. There was no moon, just starlight. A small herd of impalas stood off to the right, barely visible through binoculars. No one knew where the leopard was. It was an eerie moment, the guides whispering, everyone waiting for something to die.

Suddenly, a commotion broke out on the right. The spotlight blinked on and caught the spectacle of the leopard up a tree leap-

ing from one branch to another as four or five barn owls screamed
Chee-chee-ee and tumbled away in a panic of flashing white under
wings. The impalas barked in alarm, a sound like a short, wet
sneeze. A couple of terrified baboons yelled *Wah! Wah!* And every
human in earshot exhaled.

Our fear in the presence of a leopard was entirely natural.
We have an old and tangled relationship: Because leopards frequently stash an impala or other prey up a tree in plain sight,
archeologists think our early hominid ancestors may have scavenged for food by raiding the leopard's larder. Leopards no doubt
returned the favor. At the Transvaal Museum in Pretoria, South
Africa, for instance, there's a 2-million-year-old hominid skull
with puncture marks in the forehead precisely the breadth of a
leopard's bite.

Our relationship is, if anything, even more tangled today.
Leopards live like phantoms not just in deserts and jungles but
even within the limits of major cities. Not long ago, biologists
trapped 10 leopards in a single park on the outskirts of Dar es
Salaam, capital of Tanzania. In Nairobi, Kenya, population 3 million, I recently visited a prison compound where a leopard had
slipped in and plucked a guard dog off its chain. By stealth and
wit, leopards have managed, alone among big cats, to thrive in
modern Africa. They are invisible, and they are everywhere.

The Language of Footprints

At a remote airstrip in the arid pink hills of Namibia, on Africa's
southwestern edge, I watched one day as a Cessna 206 touched
down and rolled slowly to a stop. Two Bushmen trackers climbed
out. They were slight, mild men with small ears and prominent
rounded cheekbones. Their hair was bunched in tight little naps.
Each of them carried a sashlike hunting bag made from steenbok

skin, worn soft with heavy use. Each bag contained a small bow strung with sinew, and a quiver made from PVC pipe.

The pilot, a Namibian leopard biologist named Flip Stander, introduced them to me. This was a chore. The language of the !Kung San, as the Bushmen are properly known, uses four distinct clicks, beginning at the front teeth and moving toward the back of the throat, for which English has no equivalent in spelling or pronunciation. Tkui (pronounced with a dental click, like tht-Kooey) was the younger of the two, about 35, and sharper-eyed. Txoma (pronounced with a midpalate click, like tkk-Tcoma) was about 10 years older, slower, and more judicious. As trackers, they were fluent not just in the !Kung language of clicks and murmurs but also in the subtle and more ancient language of footprints.

We spent the next few days driving through the dry, hilly countryside in Stander's 17-year-old truck, with Tkui sitting on a plank above the left headlight to scan the dusty road for footprints. We drove in second gear, at 10 or 12 miles an hour, and mostly saw the usual litter of antelope tracks. Then Tkui held up his hand, and we got out to examine some leopard prints. They belonged to a female we'd seen the day before.

"How can you tell?" I asked.

Tkui looked at me. "Can't you see?" he said.

Stander, who was translating, tried to explain. "It's a bit like seeing people walking on the horizon and you know that one's your wife. But could you say how? It's the way she walks, but how would you describe it?"

I contemplated the inscrutable indentations in the sand and pressed him to be a bit more specific. "When asked, they'll say it's a variation in the pads. But it's also a variation in the way of walking." I made him ask anyway, and Tkui tapped his right knee and then gestured with the closed knuckle of his hand as if taking steps. "The angle of it," Stander explained. The leopard's right hind foot splayed slightly outward.

"The most difficult thing about tracking is individual identification," Stander told me. "There's nothing else that requires so much skill and intelligence." He once tested a team of four San trackers for a biological study and found they could identify individual animals (including humans) by their footprints 93.8 percent of the time. Once, when he was living out in the bush, Stander said, some canned food disappeared from his hut. The local women took one look at the footprints and named a teenager, who, on being tracked down some 20 miles away, admitted his guilt as if they had him by his DNA.

This habit of reading the world through the language of footprints is deeply ingrained. One night, an African wildcat stole some meat Tkui and Txoma had stored in a tree. The next morning, as if it were the most logical thing in the world, the two men followed the trail to the cat's lair and took the meat back. Txoma later told me that when a toddler becomes bored, his elders set him to following the trail of an ant. Tracking is simply what the San do.

Field Notes

Stander, who was 38 at the time of my visit, earned his biology degrees at Cornell and Cambridge universities. But much of his higher education had taken place over the previous 10 years learning to track with the San. In the city of Windhoek, where he worked as the official advocate for leopards, lions, and other large carnivores with Namibia's Ministry of Environment and Tourism, he dressed like a trim, bearded park service bureaucrat. Outside the city, appearances plainly didn't count. He wore a grimy down jacket, held together with bloodstains and duct tape, and a pair of shorts, which revealed that he had scribbled field notes up both legs from knee to cuff. (A bad habit, he said, picked up trying to fly a plane and record sightings at the same time.) He went barefoot everywhere, even on slopes of sharp talus. When he needed to climb a tree, which was often, he grabbed a couple of branches and then walked with almost prehensile feet up the vertical bark.

The tree climbing was essential for wiring up hunks of springbok and gemsbok as leopard bait. It was grisly work. The blood and guts got set aside to be added to a barrel of gore that had been maturing for more than a year and eight months under a tree back at camp. The trackers used it, ladled out with an old soda can wired to a stick, to make scent trails to lure leopards to bait trees and traps. "It becomes like a perfume," said Stander, both hands fanning the stench up to his connoisseur's nostrils. "The animals don't walk past it. They always stop right there."

The point of all this work was to determine how many leopards were living on a 32,000-hectare (79,000-acre) ranch called Hobatere Lodge and radio-collar them for long-term study. With the trackers, Stander had taken the conventional hunting trick of identifying footprints in the roadbed and turned it into a mathematical technique for estimating the number of leopards per 100 square kilometers of countryside. The trackers could also determine with amazing precision—98 percent accuracy in Stander's study—what each leopard had been doing the night before.

One day, Tkui headed out into the bush, with Txoma walking roughly parallel to him on the road, in pursuit of a big male leopard. "If we're lucky, we'll get it," said Stander. "It's heading straight for the bait."

The trail crossed rough ground, and the two trackers walked thoughtfully, eyes down, hands folded behind their backs. Sometimes they shared a length of copper pipe stuffed with tobacco,

too hot for lesser mortals to handle, an upgrade from the traditional pipe made from a steenbok femur. They lost the trail for a
time, and fanned out like hounds casting for a scent. Then Tkui
whistled and swept his arm to indicate the way ahead, up a dusty
riverbed. They crossed the trail of two leopards coming the other
way, a mother and her cub perhaps. Stander and the two trackers crisscrossed the trail and traded ideas. The back-and-forth
had a slow, cumulative quality, which Stander later described
as "trancelike." He broke away reluctantly and indicated some
tracks. "This is the track of the male coming over. The two leopards going the other way are probably from the night before."

"How can you tell the age of the prints?" I asked.

"By the way the wind has eroded them," he said, "and by the
superimposed footprints of ants, mice, stuff that moves around."

We followed the deep prints of the male off through the
mopane scrub. "The leopard's just gotten the scent," Tkui told us.
He could see it in the sharpness of the toe prints. After a few feet
more, Tkui crouched, sphinxlike, and peered ahead, mimicking
the leopard. The cat had dropped onto its belly, and its forelegs
had left long parallel lines in the sand, with the matching lines of
the rear legs just behind. "He's looking."

Even I could see the evidence of stalking now, in the way each
hind foot had been drawn up slowly and placed almost on top of
the forefoot. Then the trail vanished again, and reappeared not
at the bait tree but over in the shadow of a hill. The shift made
the trackers think the leopard had crossed this way at first light:
"When it's pitch dark, a leopard moves in the open. When it's
lighter, he favors darker sides, shadowy bits." We crossed scabby
bloodstains in the sand, and Tkui touched his nose to indicate
the scent trail. But the leopard had walked right past, intend-
ing perhaps to wait out the day in some shady spot on the hill.
We followed the trail up into a ravine, winding among the rocks

and scrubby vegetation, and finally Txoma said, "We'll leave him now." They had the patience of hunters. The bait tree would do its work, and waiting underneath was a box trap with a hunk of meat inside and its trapdoor propped open.

Stumbling on a Leopard

One day, we caught a female leopard, which hissed and lunged at us in her cage. With his bow, Txoma shot an arrow tipped with a syringeful of sedative into her flanks. When the leopard went down, Stander pulled her from the cage, wrapped his arms around her chest, and lifted her off the ground. The leopard raised her head and looked around with dull, puzzled eyes. Her tongue lolled out endlessly, curled round and licked her nose. She stayed in this state of twilight waking while Stander weighed her and took blood and hair samples. Then he peeled back her lips to measure her canines with a caliper. (The uppers were both about 27 millimeters long, which translates, in U.S. standard measurement, to really big.) "Is she waking up?" someone asked. "Yeah," Stander replied. He put a radio collar on her and carried her back to the cage to recover overnight.

The trackers had already identified 13 different leopards along this unpromising 12-mile stretch of road, and it reminded Stander of the estimate by scientists at the World Conservation Union that perhaps 300,000 leopards survive in sub-Saharan Africa. (In the Middle East, Russia, and much of Asia, on the other hand, leopards are doing poorly; India, with an estimated 7,000 leopards, is their biggest stronghold outside Africa.) The population naturally varies with habitat. In the rain forest, a male leopard might get by in a home range of only about 50 square miles; in the Kalahari Desert, he might need around 1,300. Either way, the average person could practically step on a leopard and never know it.

"On 1,200 occasions," said Stander, "we've walked right up to a leopard. We knew it was there because of the radio collar. And in those 1,200 times, we actually saw leopards only twice. They just move off. We hear the signals all around us." One time, Stander was flying low, radio-tracking from an ultralight. "I saw the leopard directly below me under a bush, and I told the trackers where it was. Six trackers walking straight toward the leopard. I was getting scared because they were so close. These guys were about 50 feet away. This was flat country, no rocks, just scrubby vegetation. And the leopard just slinked down on its belly, it just snaked around them, and it was on their trail. They never saw it."

Leopards generally run away from humans. Tkui and Txoma counted on it so completely, they said one day as we sat around a campfire, that they sometimes followed fresh tracks to steal a leopard's kill. The desert was harsh, and hunting was always uncertain. "If we can see the possibility of food, we follow," said Tkui, who was busy frying bread cakes in oil. So it wasn't just early hominids that scavenged for a living at the leopard's expense.

Leopards are, after all, better hunters. For instance, the steenbok, a small, generally solitary antelope, is extraordinarily vigilant against predators. A gifted San hunter can approach undetected, "if everything is perfect," no closer than about 60 feet, said Stander. A leopard, on the other hand, can often get as close as 3 feet. It doesn't need to get that close, since it can pounce 20 feet in the air, "but the moment you pounce, the animal hears you and it's off. So the closer you can get, the better." Once it kills an animal, the leopard will typically stash it up a tree and return three or four nights in a row to feed. It will often wait out the day nearby. So the idea of walking up and stealing the kill in broad daylight sounded risky. But the leopard, said Txoma, "always just runs."

Or almost always. Txoma told a story about a neighbor named Tollie, who depended on a pack of hunting dogs because he was deaf. One day, while Tollie busied himself at a pond, his dogs ran around to the other side, where they cornered a leopard in the bush and launched a horrific brawl. Tollie, hearing nothing, wandered over into the middle of it. The leopard promptly leaped up and bit him on the face and sank its claws into his shoulder. But as he started to fall, Tollie drove his spear into the leopard's chest and killed it. Then he stood up, "bleeding like a pig," and walked 10 kilometers to a village for help. Stander's team of trackers happened at that moment to be radio-tracking the leopard, which was one of their main study animals.

"So here we are tracking," said Stander, "and we find this leopard lying with a spear through his chest. The San looked around and said, 'These are Tollie's tracks,' and they reconstructed the whole thing. By the time the village people got to us, we knew exactly what had happened."

"Struggle 1 Meter"

Stander had been hoping to walk me through such a reconstruction at the site of a recent kill, and that afternoon we hiked up onto a ridge where a leopard had been hanging out. Stander had gotten a glimpse of the radio-collared animal that morning from his airplane.

The ridge was dismal for tracking, all granite pavement and pebbles. But Tkui quickly found footprints. "When the plane went over, the leopard went this way," he said. "She was running." He showed me the depth of the rear prints, and the way the claws had scooped them out in back. Then he pointed to a print where the foot had clearly pivoted in place. "It's stopping here and looking at the plane," he said.

Tkui and Txoma lost the trail for a while, so they separated and wandered ahead to where they thought they might pick it up again. "Tracking is so fundamental in hunting that it must have been extremely important in shaping how our minds work," Stander was saying. "Those who tracked best got more food. There must have been tremendous selection pressure. Watching these guys work, that makes so much sense. What they're doing here is a lot like science. You form a hypothesis about where the animal might have gone, and then you modify the hypothesis."

The two trackers meandered down the length of the ridge, weaving back and forth, sorting out several days' worth of leopard and hyena tracks, the circumstantial evidence of a recent kill. Finally, after about a mile, Tkui stopped, lifted his arm, and pointed. A baby giraffe, perhaps one month old, lay with its legs splayed, its head twisted to one side, flies circling its open belly, and a paling of broken ribs sticking up on one side.

The two trackers studied the terrain. They traded thoughts on the significance of assorted inscrutable indentations. After they'd reached a consensus, Tkui walked us through the killing. It was three nights ago. The giraffe was lying on top of the ridge sleeping. Right here. The leopard was coming up the slope. Over here. Then it caught the scent. Tkui pointed out the pause in the footprints, and the signs of stalking. "The stopping may have been a half hour, and the stalking may have been a long time too," Stander translated. "We have no way of knowing." Tkui reenacted the scene, crouching and slowly drawing his feet up between his hands until, after a distance of 80 feet, he came to a spot where there were two deep divots in the soil. It was just 16 feet from the sleeping giraffe. Tkui pawed the earth, digging in like a sprinter on the starting blocks. Then he pounced.

"The leopard landed on its back feet," he said, "and the front feet hit the animal here." He showed us the claw marks on the giraffe's left flank. The leopard lunged for its victim's throat. "As it grabbed on, the giraffe turned away." He showed us the sweeping, circling marks in the sand. "There was this big struggle. The leopard's back feet dragging, dragging, pulling this thing down." The victim might have weighed 220 pounds, the leopard no more than 90. But the contest was brief. With the leopard locked on its throat, the giraffe kicked futilely and then suffocated.

Stander jotted a few notes on his right knee: "Giraffe male, 1–2 months. Stalk distance 24 meters. Prey distance 5 meters. Struggle 1 meter." It was a terse summary of the terrible scene at our feet. Then we turned and headed back down the ridge to our vehicle.

Our Leopard Heritage

Over the course of a few days, we had begun to know the local leopards almost by personality. But I realized with a start that, except for the one animal we'd trapped, we hadn't actually seen a single leopard in the flesh. They were living undetected all around, leaving their stories in the sand to be interpreted only by Tkui, Txoma, Flip Stander, and a few other masters of the language.

It seemed to me that leopards had left their footprints not just on the African earth but also in our own genetic heritage. One time in Botswana's Okavango Delta, I went out with a friend named Dave Hamman to inspect a sausage pod tree. It was a big old monster of a tree, 8 feet across at the bottom and maybe 40 feet tall. It was just the shape that humans find innately appealing, anthropologists have suggested, because its great, spreading branches would have provided our simian ancestors with some refuge from leopards. It's what we really mean when we talk about a picturesque tree. The sausage pod tree was also a species monkeys loved to visit to feed on its huge seedpods. But this tree contained a surprise: the trunk divided about 6 feet off the ground, and there was a deep hollow just below the split, with a worn-smooth lower lip, and a sort of knob like a stoop just outside. A leopard was said to be using the hollow as a den for her cubs.

We found the leopard a half mile down the road. She lay in the shade, panting. With binoculars I looked into her vast khaki-colored eyes, and the crimson veins at the edge of the cornea seemed to swell and darken as she stared back. She closed her eyes again, and her head inclined. She moved off after a bit, and when she stepped into the tall grass, she vanished, her yellow fur blending perfectly with the dry grass blades. Her black spots disappeared among the dark spaces where the blades of grass crossed one another close to the ground.

Hamman figured the leopard was heading homeward, and we drove ahead to wait at the sausage pod tree. We passed a half hour or so trading leopard lore. I told him about a study in Barbados of monkeys that had not actually seen a leopard since their ancestors were imported from West Africa in the mid-sixteenth century. But when a researcher presented them with a football-sized object covered in a leopard print, the monkeys flipped out. The signal was apparently embedded in their genes. Richard Coss at the University of California, Davis, has suggested that a similar genetic heritage may be the reason kings, tribal leaders, and fashionable women are inclined to favor leopard-print clothing: leopards still make us a little gaga.

Just then, the female showed up. She paused and snapped at a fly that was bothering her. Then, as she approached the sausage pod tree, she made a soft coughing sound, to let the cubs within know mama was home. She leaped up onto the knob of the tree. She peered into the den. Then she squeezed her way down into the narrow opening. Her tail lingered outside. But after a moment, it vanished, too, and the tree became just a tree again, betraying no hint of the fierce life huddled up in its heart.

Every Ant on Earth at Your Fingertips

*I*n the bleak hour before dawn in a provincial capital in southeastern Madagascar, biologist Brian Fisher and a team of five field assistants stand outside the railroad station, a grand, but at the moment lifeless, French colonial structure. Bad news: The passenger train, also of colonial vintage, will not be running today. When the first railroad workers arrive, Fisher consults with them in French and Malagasy, pointing on a topographic map to a roadless site he hopes to reach. A few hours later, suitable friendships having been formed, a freight train squeals to a stop at a rendezvous outside of town, a plume of steam drifting back from the locomotive. Fisher and his crew pile tents, machetes, headlamps, mesh sacks, pan traps, cookware, and a basket of live chickens into an open boxcar, which carries them *thumpeta-thumpeta* out into the hills. Or rather, *THUMPeta-THUMPeta*. These battered old boxcars have no springs. Hanging out the open door, Fisher cannot help grinning, jumping a freight train being a forgotten childhood fantasy. For anybody else, this would qualify as an adventure.

And for Fisher? Among other hazards of tropical biology, he has endured a leech up his nose and nematodes "moving fast" under his skin. He has suffered leishmaniasis—and treatment with the heavy metal antimony. Once, in Gabon, he collapsed with malaria. A Bakwele Pygmy woman saved his life by carrying him 18 miles on her back to get an injection. He has posed as a Frenchman with Muslim moneychangers who had just learned about the U.S. invasion of Iraq. And he has been caught in the middle of a war in the Central African Republic. To get home, he had to cross a river between countries. Border guards picked him up, but he got away from them and joined up with a missionary pilot. Then he persuaded Air France to let him board a flight to Paris without travel documents or a ticket. As he negotiated this feat, he was dressed in blood-smeared clothes from a month in the field, and he was using a snake stick as a cane because one unshod foot was grossly swollen with elephantiasis. In his other hand, he carried a greasy paper bag of barbecued goat meat purchased from a street vendor because he was starving.

"I'm glad all my work is in Madagascar now," he says mildly, "because the African work was slowly killing me."

Most people would not risk such misery for a small fortune. They wouldn't do it for a reality television show.

Fisher does it for ants.

A slight, bearded researcher at the California Academy of Sciences, he has an unstoppable, infectious enthusiasm for ant biology and behavior. He is convinced that if people knew what they were looking at—if they had access to basic information about the species living in their backyards, not to mention those at the opposite ends of the earth—they would be as gung ho about ants as he is.

To help make that happen, Fisher has created a Web site

aimed at doing for ants what bird-watching guides have done for birds. So far, AntWeb (www.antweb.org) concentrates on just two regions, North America and Madagascar. Fisher's ambition over the next few years is to take Madagascar's ant fauna from the least known to the best known in the world. Then he'll use it as a model to conquer the planet, with all of the estimated 22,000 ant species on Earth identified and online. Given that only about 12,000 ant species have been described since Linnaeus named his first ant in 1758, this is no small goal.

To achieve it, Fisher has become a leader of the movement to change the way scientists collect and describe new species, using computer technology and DNA sequencing. In truth, he will use any tool it takes to get the job done faster, cheaper, more efficiently. One time, he went to Costa Rica with a vacuum cleaner, sucked up 500,000 army ants, and brought them home alive to the California Academy of Sciences, where he had a vast looping, floor-level display case winding through the exhibition area, so kids could position themselves and watch the ants come thundering down on their prey.

E. O. Wilson, the godfather of ant biology and of the conservation movement, calls Fisher's methods "industrial-strength taxonomy." He means it as high praise. Fisher himself says he aspires to the time-and-efficiency thinking of a car manufacturer.

Getting to the Good Stuff

One morning, Fisher kneels on a hillside in the rain, eyes glinting, machete embedded in the soil beside him. His headlamp lights up a patch of tiny white blobs on the base of a tree. The blobs are insects called coccids, which sip sugar from the tree and excrete a droplet of honeydew. *Camponotus* ants harvest the honeydew for food and tend the coccids like sheep.

"The ants invented herding," says Fisher. "If something were to come in and try to eat the coccids, the ants would pick them up and carry them below to protect them." *Camponotus* queens flying off to found a colony sometimes carry coccids with them to help set up housekeeping.

The conversation jumps to the genus *Melissotarsus*, which drills a hole in a latex-producing tree and lives in a silk-lined tunnel inside. "It's the only ant genus where the adults produce silk. The whole gigantic head is a big silk gland. The front legs have been modified into silk brushes, to pull out silk and stretch it to where it's needed. The funniest thing about this ant is that its middle legs go up instead of down, because it lives in tunnels. Put it in your hand and it can't walk."

It's the sort of weird behavior that makes Fisher love ants, and especially the ants in Madagascar, where he has collected more than 800 new species. Madagascar is known for its extreme diversity: "There's a place in the south," he says, "where you can stand with one foot in the rain forest being sucked on by land leeches, and the other foot in spiny bush thicket with baobab trees and a 12-month dry season."

Fisher decided to focus his research in Madagascar in 1992. To make sense of the place in a systematic way, he overlaid geologic maps with bioclimatic maps and listed 130 sites where the combined features promised to harbor interesting ants.

Says E. O. Wilson: "It's not the old style, which I admit I followed most of my life, of going where you can get. Brian laid out a series of sites that he felt were necessary for a complete sampling and he just got there. That's an improvement."

Fisher and his crew generally travel from site to site in a Toyota Land Cruiser, staying in the field for two months straight in the thick of the rainy season, the best time for ants and the worst time for roads. At one point, when the vehicle got bogged down in an area with no trees for attaching a winch cable, Fisher and his team hiked 6 miles (about 10 kilometers) cut down some trees, carried them back, dug holes, planted the trees, winched the vehicle 30 feet (10 meters) forward, then dug up the trees, replanted them, and did it again and again for two days.

Among other techniques for crossing flooded rivers, he sometimes attaches ropes to the sides of the Toyota and positions swimmers midstream. Then he drives in till the water laps the windshield. (Any deeper, says Fisher, is risking a rollover.) He shuts down just before the engine floods and waits while the swimmers haul the ropes to the opposite bank and tow. It doesn't always work.

One time, the engine died at night in the middle of a fast-moving river. The water rose. Fisher walked to the nearest village and woke up the headman, who organized a caravan of oxcarts hauled by zebus, the local cattle. The party arrived back at the river at dawn, and the headman stood on the hood of the car crying out orders to the five zebu teams spread out ahead. The car started to move. But the opening in the riverbank was too steep and narrow for the zebus. So Fisher walked 15 miles (about 24 kilometers) to a shrimp farm, and then bushwhacked back ahead of a borrowed tractor.

"You read these stories about cowboys going all night, and I'm thinking, 'How do they do that?' But you can, when the time comes. We didn't eat for two days, and you can do that, too."

Protocols for Collecting

Doing it for ants does not strike Fisher as odd. He grew up in Normal, Illinois, the son of a science education professor and a grade-school teacher, both tolerant of his youthful interest in animals (though his mother screamed when a pet alligator joined her in the shower). He came to science intent on discovering new things. As a graduate student, he studied the ways ants and plants benefit one another. But on a research trip up the highest mountain in Panama, it dawned on him that all the plants were already named. "I said, 'What about these ants?' And nobody knew the species, or even the genus." So he spent a year getting retooled in insect systematics. Then, for his PhD thesis, Fisher set out to retool the science he had just learned, testing and revamping techniques at every stage of the taxonomic game.

"These are the magic devices," Fisher announces one morning, as he hangs up lanternlike cloth bags on a clothesline under a tarp. "This is why we are able to go to a place and collect all these things that none of the other collectors saw." A beetle researcher originally developed the device, called a Winkler bag, but Fisher refined it and established the protocol to make it a standard tool for biological inventories.

Here is the protocol: Get down on your knees, machete in hand, and start hacking at the patch of earth directly in front of you. Shovel the resulting hash into an upside-down witch's hat, with a sieve near the brim. Then bounce up and down repeatedly. The soil, which shakes through the sieve to the point of the witch's hat, doesn't look like much.

"You'd never guess that in that brown mess, there are all these beautiful brown insects," says Fisher. "And when you're doing it, you're thinking there's nothing here."

The soil gets carefully measured into mesh sacks and hung up inside a Winkler bag. The Winkler exploits the tendency of soil-dwellers to escape danger by heading down, enticing them into a plastic bag filled with alcohol. Next day, the alcohol has become a tea-colored soup studded with worms, shrimplike springtails, and drifting ants, their delicate legs akimbo.

At the same time, yellow pan traps and malaise traps are doing their dark work. The pan traps are disposable plastic cereal bowls filled with soapy water. Ants and other creatures cannot resist climbing in for a swim. Malaise traps are tentlike mesh devices set up along natural flight paths. Flying insects escape danger by moving up, not down. So when insects land on the mesh, the angled ridge of the trap leads them up and over the edge into a jar of alcohol at the top. More soup.

"I like traps that work while we sleep," Fisher says.

Shortly after dawn, he and his crew fan out across a ridge. Like any cost-driven entrepreneur, he has opted to send much of the basic work of taxonomy offshore, where he can train and pay five technicians for the cost of a single technician in the United States. "You have to move to where the labor market can support the labor," he says. "And it's also where the biodiversity is, and where you want people to learn about it and appreciate it."

The crew, including a former taxi driver, a cook, and a mosquito biologist, have become zealous ant hunters. (Some past crew members have gone on to achieve advanced degrees in entomology.) The forest soon resonates with the splintering of rotten logs and the snapping of branches. One worker wraps his fingers under a mat of bamboo grass and peels it back, the muted popping of tendrils like a string of firecrackers going off in the distance. Then comes the soft *pttt-pttt* of a "pooter," a pencil-thick rubber hose held between the lips and used to suck up ants. It has a fine-mesh screen to divert specimens away from the mouth and into a collecting tube. Off in the distance, someone is flailing away at the underbrush with a stick, collecting everything that falls out on a white square of cloth stretched over a PVC frame, and then—*pttt-pttt*—sucking up the interesting stuff.

The workers stop chasing ants only for their ritual food break every three hours. Even then, they hold up tiny vials to a high window of light in the forest canopy, and recite treasures for Fisher to scribble down in his "Rite in the Rain" notebook:

No More
Eviscerated Giraffes

Like Fisher's technicians, I came late to natural history. I started out as a writer of obits, at the age of 18, at a small newspaper in New Jersey. It was the standard apprenticeship then, because it taught you to spell peoples' names right. Otherwise, you had to deal with sobbing widows on the telephone, sometimes followed by hostile sons threatening to punch you in your stupid little face. It was the summer of Woodstock, but I spent my time chatting with oily funeral home directors and deploying the language of "visitation hours" and "in lieu of flowers"—all translated then into lead type on big clanking Linotype machines. One night I got a call from a family member who ignored my questions and simply dictated his mother's obituary. Finally I said, "Look, buddy, who's writing this thing, you or me?" I learned the next day that he'd been the editor of the paper for the past 40 years. This may suggest why I've never held a full-time job for more than 2 years in my life.

My training as a writer (and a diplomat) began long before I became an obits man. My father was a college writing teacher, whose mantra was "Rewrite, rewrite, rewrite." He tended to mark up his students' papers with comments like "PV! NDG! RNF!" which stood for "PASSIVE VOICE! NO DAMNED GOOD! READ NO FURTHER!" I think he was slightly less blunt with family members, but maybe not. I have visceral memories of broad slashing strokes of red ink. My father also introduced me to The Elements of Style by William Strunk Jr. and E. B. White, which became my handbook as a writer: Keep it short. Use nouns and verbs, not adjectives and adverbs. Avoid the passive voice. Let style arise from plainness and simplicity. Rewrite, rewrite, rewrite.

It was a long time before I figured out that I liked writing about animals. In college, I visited the section of the campus called Science Hill just twice, first as an antiwar protester, and later for my job as a projectionist. After studying poetry (Spenser to Milton) in college, I went back to a bigger newspaper in New Jersey, where I covered local politics and colorful murders, including one committed with a bowling ball drill. (You work with what you've got.) One day when I'd been scooped by another newspaper on a house fire, I was desperately seeking a follow-up story. Just before deadline, I phoned my editor, "Good news, Jim, the baby died." Soon after, I turned in my resignation. A few months after that, when I was starting to spread the peanut butter sandwiches extra thin, a magazine editor agreed to let me write about New Jersey's state bird, the salt marsh mosquito, and I was hooked.

Since then I have held two other jobs, one as a staff writer at a short-lived science magazine, and later as the managing editor of Geo magazine. Geo was trying to recover from a period of grim and graphic German ownership, and the new motto on the cover was "Travel and Discover the World," though among ourselves we thought "No More Eviscerated Giraffes" was more like it. It was a wonderful place to work, dispatching me to Panama to write about three-toed sloths (a reader commented that their sex lives reminded her of certain men she had known). It also sent me to Jersey (that would be Old Jersey) in the Channel Islands to interview Gerald Durrell, the zookeeper and author of the classic My Family and Other Animals. The interview started at 10 a.m., with glasses of whiskey the size of swimming pools, and Durrell waxing romantic on the possibilities of being reincarnated as a fur seal: "One would swim a thousand miles to marry a female fur seal! They look as if they've been made out of molten gold." He reminded me that it was possible to write about the natural word—even the occasional eviscerated giraffe—and still have fun. Geo eventually died, and I decided to follow Durrell's advice. As a full-time writer, I have lived at the mercy of editors (and other animals) ever since.

"*Tetraponera*, low vegetation. *Anochetus*, under log. *Pheidole*, ex rotten log." When they have finished with this patch of forest, it looks like the aftermath of a vigorous police search, everything upside down and inside out. It will take the ants a week or two to put it right again.

Assembly-Line Taxonomy

Back in Antananarivo, the capital of Madagascar, the field crew and four other parataxonomists sit shoulder to shoulder at microscopes sorting the results of their collecting. It takes one person a full day to separate a kitchen plateful of mixed specimens by order, and then another four hours to sort the ants into 40 or so genera. The Malagasy technicians also handle the maddeningly fussy work of gluing each ant to the tip of a tiny triangular piece of cardboard, to be mounted on a pin and arranged in a box. In his entrepreneurial mode, Fisher once ran a time-budget analysis, which revealed that his staff was spending 25 percent of their effort on making and verifying labels. "I said, 'This is not scalable. This is not the method.'" So he devised a streamlined system of mass-produced labels. The assembly-line approach gets a collection ready for him to cart home in months, instead of years.

Why the hurry? In the venerable tradition of Victorian naturalists, biologists used to go out into the field for a few years in their youth, and then retreat to the dusty back rooms of museums and universities to spend the rest of their lives cataloguing their treasures. But with habitats and species disappearing faster than scientists can describe them, Fisher says the glacial pace of traditional taxonomy makes no sense.

His solution is to democratize the science of biological classification. AntWeb makes detailed photographs of museum specimens available at the click of a mouse. It also provides links to the scientific literature for each species at an American Museum of Natural History Web site (antbase.org). Thus the Internet provides anyone anywhere with the tools to discover new species. In "team taxonomy," Fisher says, backyard naturalists will be able to recognize what's unusual in their own neighborhoods and then collaborate with professionals to enter it into the annals of science.

The problem with traditional methods of collecting and preserving specimens isn't just that they're slow, Fisher says, but that they often make the results inaccessible to almost everyone. Everything comes down to the specimen on its pin, a sort of sacred totem to the taxonomic priesthood. The shaft of the pin holds tiny, folded scraps of paper full of spidery notations of crucial data like the time and place of collection. If the collector was especially enthusiastic, or if the species has been renamed in the course of a taxonomic revision, a pin may hold up to 10 different labels. To read the bits of paper, you have to slide them off the pin with a pair of tweezers.

That is, assuming you can find the right pin. The rules dictate that there should be a "type specimen" establishing the characteristics of the species for all time. But as he traveled through the world's great natural-history museums trying to assemble the basic data about Madagascar species, Fisher found that past collectors often failed to label specimens as "types." They also neglected to name the museum where they were stored.

Idiosyncratic organizing schemes further complicated Fisher's search. Specimen pins typically get stored standing upright in neat rows inside glass-topped boxes. One national museum boasted more than 300 boxes of ants. "But if you wanted to find collections for Madagascar," says Fisher, "you had to start at box one and study every box." He also had to use his field headlamp to see the ants in the dim museum light.

"It was abysmal. It was sad," he says. "The myth is that this is forever. You create the specimens, and that's your legacy forever. You think it will be looked after by society." He pauses. "They've been forgotten. They're shoved in a corner nobody's looked at for 100 years. They're in such disarray!"

The emphasis, says Fisher, shouldn't be on the specimen as the sole bearer of all information. It should be on the Web site or the database, with the specimen serving only as a voucher. "It's such a subtle change, but it really frees you up," he says. The Web gives a collector the space to record detailed ecological data about a specimen. Other people can find the information almost instantly, and organize it any way they want—by "Madagascar type specimens," for instance, or by "ants worldwide" found "ex rotten log."

Web sites can also include genetic information. Proponents of DNA barcoding analyze a short piece of the genome—about 600 nucleotide base pairs out of the millions or billions in an organism—to see whether, say, two very different-looking individuals might in fact be the same species. To make that determination by traditional means, you need a taxonomist like Fisher with years of postgraduate training. In the case of ants, he must collect about 20 specimens, sit down at a microscope, and make 15 measurements on each one, where differences of a millimeter are crucial. But for $1.50, anyone can get a sequencing machine to look at a single mitochondrial gene and say, "This might be a new species." On his last collecting trip, Fisher shipped off the genetic raw material and in three weeks got back all the DNA sequencing data.

"It told you immediately what site was most diverse and what site was most unique," he says. "Taxonomists would shoot me if I said, 'Genes could replace taxonomy,' but for a lot of species, genes are the only data we're going to have. We have buckets of flies and wasps that nobody's working on. If we sequence them all, we'd at least have an idea what's out there. People say, 'That's not taxonomy.' Well, it's a type of taxonomy. It allows you to do all sorts of stuff about endemism, suggest new species, how different and how similar things are, without ever knowing the names of species.

"What they're afraid of is taxonomy being done poorly. What I want to show is that new tools will allow more people to participate in taxonomy, and we can do better taxonomy." Instead of being buried under a hundred lifetimes of unsorted specimens, taxonomists will be free to cut straight to the difficult questions where their expertise can make a real difference. It isn't the end of taxonomy, says Fisher. It's the golden age.

Hunting for Treasure

Fisher and his group move to a new site called Vatovavy, a fragment of forest in a saddle of land between two peaks. It has survived deforestation because it is sacred to the dead, who are entombed in a building at the edge of the woods. Fisher is soon on his knees in front of a log crawling with *Cerapachys* ants.

"They're so beautiful to watch. They don't have a trail pheromone, so they walk head to butt, head to butt. They're like little beads on a necklace walking around, little shiny black pearls." *Cerapachys* is a relict genus, dating back 100 million years to when Madagascar split away from the mainland. Like a lot of the island fauna, says Fisher, it opens a window on early evolution. Elsewhere, *Cerapachys*-like ants evolved into army ants; it probably happened more than 90 million years ago, before Africa and South America separated. But here, the ancestral lineage still thrives. "They feed off the larvae and pupae of other ants. They trek out of the nest and into another ant nest and steal all their larvae and take 'em home and eat 'em. You can tell they're primitive the way they walk around. They can't even scurry. They hadn't *invented* scurrying yet."

To Fisher, every ant, like *Cerapachys*, tells a story, and the stories only make sense if taxonomy can explain where a species fits in the scheme of evolution. Understanding the story isn't just a question of aesthetic or intellectual interest.

"Imagine the storehouse of evolutionary information," he says, "in 10,000 species living in the dark wet soil, having to keep fungus and bacteria off their young and off themselves. Ants are chemical factories. They produce antifungal and antibacterial agents. So you think, my god, bioprospectors would love to get into this." Ants could become a precious pharmaceutical resource, especially now that resistance has rendered many conventional antibiotics ineffective.

Fisher argues that ants are also valuable in determining which habitats to preserve and which to let go. "Bird people and mammal people will go into a place and say, 'There's nothing here, it's all been hunted out.' And we'll go and say, 'There's the most interesting stuff. You've got to save it.'" Ants can't readily cross rivers, mountains, and other barriers. So they stay put and become specialists in a relatively small home range. This makes them a useful tool for detecting ecological subtleties that might not be evident just from looking at larger animals. To Fisher, a varied ant population suggests that "the forest has a unique history, a unique assemblage, and should be preserved even though it doesn't have some fuzzy vertebrate living there."

He balks at the idea of ants as a mere indicator species. "They're not just a shortcut to finding diversity, they *are* the diversity." It is not unusual, for instance, when one member of the team shows up during a snack break at Vatovavy with 10 genera of ants from a few hours of collecting. "You can remove all the birds and still have a forest," says Fisher. "But you can't have a forest without invertebrates. It won't function anymore. The ants are the glue that holds it together."

He concedes that no one is likely to save a forest just for ants. But Madagascar President Marc Ravolomanana, a former businessman, has publicly committed himself to tripling the area of protected habitat on the island. Fisher is one of a group of scientists, conservationists, and government officials helping to determine which habitats merit protection. Other members of the group say his ant data are so thorough that, as they formulate their recommendations, they are using it in ways "almost comparable to the bird and mammal data."

On the drive back from Vatovavy, Fisher makes a flying stop at a conservation meeting in a national park. He bumps into Bernard Koto, the regional governor, a formidable figure with a background in conservation who is feeling pressure to come up with more acreage to be protected, and fast. (President Ravolomanana's deadline for tripling protected acreage was 2008.) Koto wants to know what Fisher has been finding at Vatovavy and whether the forest has the potential to become a reserve. It does not seem to trouble him that the biodiversity they are talking about is in the form of ants. Fisher promises Koto that he can have the Vatovavy data not in two years, but in two months. Meanwhile, his crew members are quietly folding themselves back into their vehicle.

"This," says Fisher, his foot already on the accelerator, "is how conservation gets done."

Jelly Bellies

Look at the jellyfish—look at them, that is, in their own element, underwater—and the first thing that strikes you is their improbability, that something so beautiful could be made from such unpleasant stuff. It's as if a child with a bad cold had suddenly developed a gift for blowing Fabergé eggs through the nose. Jellyfish are one of nature's dreadful little miracles, repulsive and lovely at the same time: They weave a mucuslike gel into some of the most elegant forms on Earth.

They weave it into a way of life, and not the blobby, listless life we generally ascribe to creatures of the gelatinous persuasion. Until recently scientists knew jellyfish mainly from the dismembered goo dredged up in their nets, a kind of roadkill taxonomy.

"It was like studying the rain forest by drifting over it in a balloon, flinging out a net, and seeing what you came up with," says Richard Harbison of Woods Hole Oceanographic Institution. "You might catch a hummingbird. You wouldn't know that it feeds on nectar." But over the past 30 years, by using scuba gear and submersibles, scientists have discovered that jellyfish in their own habitat are among the most numerous, effective, and, in some ways, ingenious predators on Earth. Without eyes (in most species), ears, or even a brain, jellyfish manage to affect almost everything that lives in the ocean.

One placid summer morning in Monterey Bay, on the central California coast, an unmanned submersible named *Ventana* rocked beside its mother ship, the *Point Lobos*, with nothing showing above the surface but a little plastic hot dog some joker had mounted on a radio beacon. The submersible slipped away from the ship off the starboard bow, trailing blue-white plumes from its thrusters. A thick yellow cable uncoiled in its wake, tethering it to the ship. "Clear to dive," someone said. A brief flurry of white geysers, and *Ventana* subsided with its plastic wiener into the depths.

Aboard the *Point Lobos* in the forepeak, a dark little room below the bridge, Bruce Robison of the Monterey Bay Aquarium Research Institute (MBARI) sat at a bank of video monitors, studying the images sent up by the submersible. They were stranger than any transmission from outer space. Constellations of jellies appeared out of the darkness, looming brighter and larger as *Ventana*'s camera drew close. One jelly looked like a cocktail onion, one like a rocket, one like a Rastafarian from Alpha Centauri.

A Mouthful of Sting

These jellies were actually doing stuff, not just drifting. One comb jelly, a sort of flying wing up to 2 feet across, behaved like a crop duster working a cornfield. On its trailing edge it had a comb-like row of paddles, called cilia, for propelling itself through the water. These cilia produce almost no bow wave, so comb jellies can sneak up on their prey. Scientists call it "stealth swimming." But this particular species produces plenty of turbulence to the rear, so it feeds by cruising one way for a few yards, then climbing up 3 inches and methodically reversing course to capture tiny crustaceans fleeing the wake from its first pass. I got the impression that jellies are not necessarily helpless.

"Here's *Colobonema*," said Robison, as the submersible descended. The pilot, working a joystick, moved the 6,500-pound *Ventana* toward the animal in question and then felt the submersible strain against its tether. "Gimme some slack," he called up to the bridge. "I'm like a dog trying to get to a fire hydrant."

"*Colobonema* is the one that drops its tentacles when it's pursued," Robison said. "Sometimes you just look at it funny. But usually it takes a disturbance." Dropping tentacles is a distraction technique, like the way a lizard sheds its tail in the heat of a chase, so its pursuer will lunge for the wriggling appendage and let the lizard escape. But jellyfish do it better.

Ventana took up the chase, its thrusters roaring, its lights blindingly bright. Sensing a threat, the jellyfish merely pulsed out of reach. *Ventana* made two more passes without ruffling the *Colobonema*'s jelly belly cool. "You gotta give this guy points," the pilot said. "He's holding it together. I can't hassle him any more." But then the pilot's eyes gleamed as another *Colobonema* loomed up in his sights. The jellyfish darted away—and this time left behind a bright shower of shed tentacles riddled with stingers. Escape and vengeance in a single sticky package. Robison believes the tentacles are bioluminescent, encouraging predators to lunge for them in the darkness of the ocean depths as the *Colobonema* escapes. For the predator, this could be worse than a mouthful of sting pain: Some species of jellyfish actually use bioluminescent displays to light up a predator and make it a target for everybody else. "Nature's wonders," said Robison, eyebrows raised. "Now let's get outta here." And *Ventana* sank another 300 feet into the abyss.

Robison, grizzled and garrulous, is an old-style net-dragging fish scientist who was born again midcareer into the world of jellyfish. It happened because a bird biologist named Bill Hamner had become allergic to feathers. Hamner was considering a switch to oceanography and went along with Robison on a 1969 collecting trip. "He kept looking at the stuff we dragged up in a net, and it looked terrible. So he said, 'Why don't we just go down and look at it.' A birder's mentality." Robison cringed at the thought. "I said, 'You can't do that. Nobody's ever done it before. This is how we do marine science. We tow nets.'" But Hamner persisted.

"We started diving down there," said Robison, "and it wasn't anything like we thought it was going to be. The principal difference was the gelatinous stuff. The jellyfish were so much more abundant and so much more important ecologically than the nets ever told us."

Hamner, now at UCLA, went on to introduce an entire generation of other scientists to the joys of underwater gelatinology. He is now widely regarded as the intellectual godfather of those who swim with the jellyfish. In the early 1990s Hamner also made it more appealing to put jellyfish on display, by modifying an aquarium tank called the kreisel, which uses circular currents to keep the jellies floating in midwater. Until then, the average layperson's response to jellyfish had consisted of the single word "yuck" (and a liberal salting of meat tenderizer for soothing nasty stings). But the hypnotic, even hallucinatory, beauty of these creatures was a revelation. People now buy jellyfish Beanie Babies, keep jellyfish as pets, and, in one fashionable San Francisco restaurant, sip their drinks under jellyfish chandeliers. After roughly 600 million years of unglamorous obscurity, jellyfish are suddenly cool.

A few clarifications are therefore in order: They aren't fish, and they aren't jelly either. Unfortunately, no better name has suggested itself. One scientist refers to them affectionately as "sea boogers," but this surely will not do.

The creatures we call jellyfish do not fit into any neat biological category the way, for instance, lobsters, shrimps, and crabs are all classified as crustaceans. Scientists divide the entire animal kingdom into roughly 30 groups, called phyla, and gelatinous creatures turn up in at least five of them. (By contrast, all vertebrates on land and in water, including cats, bats, birds, snakes, dolphins, and humans, make up only part of one phylum, the chordates.) Most so-called jellyfish are either ctenophores (comb jellies) or cnidarians, which typically have the conventional bell-like shape and stinging tentacles. But some worms, snails, and squid also pass for jellyfish, as do several of our kin among the chordates. An appendicularian may not look like our distant cousin. It looks, in truth, like a tiny, fluttering strip of transparent ribbon candy. But it has a nerve cord like us.

Fun with Mucus

Why have so many creatures evolved the gelatinous form? Because it's what works in the ocean. The building material is cheap and readily available: a few proteins together with a lot of polysaccharides—long sugar chains that soak up water—producing an animal that's 95 percent liquid. So a gelatinous animal can rapidly increase in size when food is plentiful, and it can spew out huge blooms of new individuals. When food becomes scarce, a jelly can shrink down again and travel light. Being, in essence, organized seawater means a jelly can manipulate its buoyancy for migrating up and down in the water column. A sturdier body isn't necessary in the ocean because there's hardly anything to run into and gravitational force is less of a factor. A more solid body would also be harder to conceal. In a world with few hiding places, being transparent is a jelly's best defense.

But the gelatinous lifestyle isn't necessarily diaphanous or flimsy. Consider, for example, a problem in jellyfish engineering: Muscles can only contract; they can't stretch themselves out. The human solution is to use opposing muscles anchored to bone to bend and unbend our limbs. But jellyfish are far too simple for that. They can contract their bodies to shoot out a jet of water, but they have no opposing muscles to yank themselves back into shape. So some jellyfish are built around a transparent disk called the mesoglea. It's like mucus but with collagen fibers mixed in to make it tough and rubbery. It has shape memory, like the padded-shoulder insert in a woman's jacket. When the jellyfish contracts its muscles, the mesoglea bends with it—and then springs the entire jellyfish back to its original shape.

With slime as your basic building material, you can apparently do almost anything. For example, an appendicularian uses mucus to construct an elaborate house, a mesh structure ranging from the size of a pea to the size of a doghouse. Suspended within this home, it merely flutters its tail to drive water through a pair of tubular circulators and into two screens, also made of mucus, which filter out minute food particles. If the house or the screens get clogged, the appendicularian simply moves out and builds a new place. A comb jelly, in turn, has evolved to feed on appendicularians. It opens up like an umbrella and engulfs the appendicularian, house and all. Then it waits patiently until the appendicularian, fluttering within, realizes it's not getting any food. "The appendicularian reacts as if its screens are clogged," says Harbison, who recently observed the behavior for the first time in the Antarctic. It leaves the house—and swims straight into the comb jelly's mouth.

Reading about jellyfish, I started to make a list called "Fifty

Clever Things to Do with Mucus." Many of the cleverest tricks have to do with killing.

Another jelly, a chain of linked individuals called a siphonophore, makes tiny fishing lures with its own flesh. They look deceptively like larval fish or small crustaceans. The siphonophore dangles these lures at the end of long tentacles, jigging them through the water with a convincing twitch. Fish that fall for them get a mouthful of deadly stingers, and their carcasses are slowly hoisted up into the jellyfish. In one case a single siphonophore was fishing 85 lures at a time, like a party boat out for bluefish.

On the video monitor *Ventana* was showing us a commonplace siphonophore named *Nanomia*, a frilly little thing 6 inches long. "By our estimate," said Robison, "that one species takes a quarter of all the krill in Monterey Bay. They're probably the dominant predator of krill here, duking it out directly with blue whales and salmon."

Another siphonophore named *Praya*, which is arguably the longest animal on Earth, lurked somewhere deep below us. MBARI researchers once announced that they had measured a *Praya* longer than a blue whale (though only about as thick as your wrist), causing anxious parents to phone in about whether they should let the kids swim with a 130-foot predator. MBARI gently advised them that their children would be in danger only if they happened to swim 800 feet deep, where *Praya* lives, and even then it would help if they resembled small crustaceans. But I had the feeling that the scientists were secretly thrilled to see jellyfish getting even this misguided respect.

Curtain of Death

Jenny Purcell, who wears a cap embroidered with a sea nettle rampant, is a leading proponent of the idea that jellyfish are among the great predators on Earth. Out on the Chesapeake Bay in Maryland one summer morning, she was raking her bare fingers through tray after tray of slime, sorting out the pearly little globs called *Nemopsis*, a cnidarian, from the thick mass of *Mnemiopsis*, a comb jelly. (Comb jellies have no sting, Purcell explained, and other jellies with stingers geared to small, soft prey may not be potent enough to penetrate human skin.) Now and then a sea nettle turned up or a fragmented moon jelly, which she pieced back together like a jellyfish jigsaw puzzle, just to make sure it was one individual and not a crowd. The idea was to survey this neck of the bay, using plankton nets and scuba gear, to see which jellyfish were at home and how they might be affecting the food chain.

People tend to regard jellyfish as a problem for two obvious reasons: We don't like getting stung, and we don't much care for getting slimed either. In the Chesapeake, sea nettles are sometimes so thick it seems as if you could walk across them; they have such a fiery sting that most people simply do not swim in the heat of summer. Blooms of jellyfish also seem to have become more common around the world. On the French Riviera a few years ago big magenta-colored *Pelagia* jellyfish turned up in such dense clumps that they shredded 5,000-pound fishing nets. In Japan jellyfish have clogged seawater intake pipes and forced a nuclear power plant to throttle down. But Purcell and other researchers have begun to think that jellyfish affect habitats—and humans—in ways that are much subtler than that, and also far more profound.

"They're always there, and always eating," said Purcell, now a researcher at the Shannon Point Marine Center in Washington. She once studied a bay in British Columbia where the dense, swirling schools of herring turn the water milk white with their spawn. But one year a bloom of crystal jellies wiped out the entire crop of larval herring. "And even when they're not decimating the fish directly," she said, "they're chowing down on food that would otherwise go to fish."

Purcell is a low-budget researcher. (All jellyfish research is low-budget, relatively speaking: The U.S. government spends roughly $14 billion a year studying outer space but just $160 million getting to know all the alien life-forms in our own oceans.) Her idea of a submersible is a video camera mounted on a rack of PVC tubing, with a $2 model airplane propeller from the hobby store to indicate the speed of the current. But she is painstaking and meticulous. One of her techniques is to strap on scuba gear, capture a jellyfish in a plastic jar, and immediately inject preservative into the jar at the jelly's natural depth, so she can measure its last undigested meal down to the slightest morsel. In one case she calculated that a single sea nettle as big around as a saucer was eating up to 18,682 copepods a day. Copepods are tiny flealike crustaceans and happen to be among the most numerous animals on Earth. Not even their mothers would miss 18,682 of them. But if they weren't being glommed by booming jellyfish populations, the copepods might be food for the kind of fish we eat. Purcell figures that in the Chesapeake Bay, sea nettles and comb jellies together may also consume up to half the daily production of bay anchovies, an important food for game fish and aquatic birds.

The world probably does not need new reasons to loathe jellyfish. So another clarification is in order: Jellyfish misbehavior may be at least partly our fault. No one knows for sure, but jellyfish blooms may occur in part because we overload a body of water with fertilizers and sewage. This leads to an increase in the planktonic plants and animals on which jellyfish feed and creates a low-oxygen environment in which fish die but jellies thrive. Jellies may also benefit when we knock out their major rivals through overfishing. Some scientists estimate that bringing back oyster populations to their old levels in the Chesapeake would reduce the jellyfish population by 90 percent.

But even now scientists have only the most rudimentary understanding of the relationship between jellies and other marine species, and our ignorance may have unfathomable consequences. In the Gulf of Maine, for example, copepods are the primary food for larval cod and other fish. But the cod have gone bust, and Marsh Youngbluth of Harbor Branch Oceanographic Institution in Florida worries that gelatinous creatures may be elbowing ahead of them at the copepod smorgasbord. "People think, *We'll just stop fishing and the cod will come back*," says Youngbluth. "Well, it doesn't always work that way." It can be a matter of who fills a niche first or more effectively. Cod eat only what they can see. But jellyfish are nonvisual predators. They simply wait, everywhere, with dangling toxic tentacles like "a curtain of death" for the copepods. In the 1980s the Black Sea fishery was suffering from pollution and overfishing. Then some ship flushed its ballast water and accidentally introduced the comb jelly *Mnemiopsis*, which blossomed in its new home. The human catch of anchovies subsequently plunged from 600,000 down to 14,000 metric tons a year, and it has yet to recover.

Jellyfish for Dinner

But if jellyfish are always out there and always eating, then surely someone is eating them too? At Woods Hole, Richard Harbison is still adding to his list of several hundred species known to practice what he calls "gelatinivory" (rhymes with "quivery"). The list includes powerful game fish like tuna, utterly ungelatinous 1,200-pound leatherback turtles, ocean sunfish the size of Volkswagens, and such oceanic birds as fulmars and phalaropes. Gelatinivory, says Harbison, may be as important an adaptation for marine animals as is eating plants. Jellies appear to be the major food source for chum salmon and butterfish, for example, and Harbison believes their dietary importance to other species like red snapper may be underestimated because the evidence gets digested so quickly.

"Jellyfish are very easy to eat," says Harbison. "They don't swim very fast. But they're mainly water. So one of the major things a predator has to figure out is how to get rid of the water." Sunfish, leatherbacks, and some other species have evolved pharyngeal "teeth" at the backs of their throats. Having swallowed a jellyfish, they regurgitate it against these teeth to strain out the water and save the edible tissue. When sea turtles mistake balloons and other plastic objects for jellyfish, this process of regurgitation may actually suffocate them. In addition, certain fish such as those in the genus *Peprilus* have evolved pharyngeal sacs, like a chicken's gizzard, lined with barbed teeth for grinding up jellies; Harbison theorizes that they may also use an enzyme to separate the edible tissue from water. "A fish will eat a lot of jellies, and you can see its stomach bulge out," he says. "Then it starts making these gulping motions with its mouth, and the belly will start to go down, and then it will eat jellies again. So what I think it's doing is expelling water." Jellyfish themselves are among the most voracious predators of other jellyfish. Sometimes you can see one wrapped up inside another, neat as a piece of fish shrink-wrapped at the local supermarket. Look closely and you may even see the victim's victim, a krill or copepod it was eating when its final meal was so rudely interrupted.

The Great White Shark of the jellyfish world is a comb jelly of the genus *Beroe*, and one day I went out in search of *Beroe* with Dave Wrobel, a jellyfish keeper at the Monterey Bay Aquarium. It was a blustery, overcast day, with the waves rolling in and fog intermittently obscuring the rocky shore of Point Piños. Wrobel cut the engine on the Boston Whaler in the middle of a long slick formed by a meeting of winds and currents. Strands of drift kelp twined in the dark green water, and we began to see jellies all around us. Wrobel and a docent named Leon began to scoop up sea angels, a gelatinous snail with small, fluttery wings and a long diaphanous mermaid bottom. Bits of siphonophores drifted past and the tough, rubbery carcass of a salp, a gelatinous chordate, studded with tubercles. I scooped up a *Beroe*, and in my hand it was shapeless, quivering like a soft candy lozenge. But when I put it in our water-filled collecting bucket, it spread out again, becoming delicate and conical, its comb-plate fringes shimmering in the light. The wind started to kick up, and we headed back to the aquarium.

I hung around to watch *Beroe* in action because I had heard that they are unlike any predator on land. A *Beroe* can engulf a victim twice its size in a single bite. Then it seals itself shut like a Ziploc bag and does a sort of hula shimmy until the excess water comes leaking out its mouth. "Mucus," one researcher had told me, "is a wonderful material." The modified cilia of the mouth also serve as teeth, making *Beroe* the only jellyfish capable of delivering a bite. I watched a *Beroe* for a while taking chunks out of another jelly, and it was like a Christmas ornament making war on a Victorian lampshade.

But then I got distracted by a sea gooseberry, which gets its name from its size and spherical shape. It was bumping busily back and forth, powered by its comb plates. Streaming behind it like Rapunzel's hair, fine tentacles caught bits of food from the passing currents. After a while the sea gooseberry stopped, as if to contemplate its quandary: Say you have no arms, and you've struck on this system of collecting food with the hair on the back of your head. Now how do you get the stuff around to your mouth? The jelly began to spin like a top, till its hairlike tentacles wrapped all the way around it. Then, languidly, it drew the tentacles one by one back through its mouth, slurping off the morsels of food. Another jellyfish caught my eye, and then another. I lingered blissfully among the jellyfish, and the word "blob" never once crossed my mind.

All Piranhas Want Is a Nice Piece of Tail (But Not Yours, Thanks)

Em rio que tem piranha, jacaré nada de costas.
(In a piranha-infested river, alligators do the backstroke.)

—BRAZILIAN SAYING

One morning in my midlife crisis, when I had considered and rejected the thrill-seeking possibilities of nude skydiving or drag-racing on a mountain road in a red Corvette convertible, it occurred to me that it might be a good idea to go for a swim in a tankful of hungry piranhas, at feeding time.

I made the necessary arrangements, and soon after I was standing in the dark in front of the piranha tank at the Dallas Aquarium, surrounded by seven-year-olds excitedly telling one another how quickly piranhas can reduce a human being to bloody stew. One of the boys was lashing his head from side to side, teeth bared. Then a couple of other boys joined in, swarming around me. It was a horrible sight.

I had seen piranhas before. But the aquarium's animals were bigger than I expected, some nearly a foot in length, and there were 40 of them in an area not much larger than a hot tub. With their lower jaws thrust forward, they were pugnacious looking creatures, like bulldogs in silver body stockings.

A curator led me to a changing room, and I put on a pair of bright red bathing trunks, just to make sure the piranhas couldn't miss me. The curator had assured me that there was no real danger. Piranhas were the innocent victims of deep-seated human paranoia about the natural world, and the purpose of my swim was to debunk these old fears. The curator and I both knew that people in South America swim with piranhas all the time and generally emerge intact.

Then we arrived at the tank and at the last minute, as if the thought had just occurred to him, the curator asked, "What if something *does* go wrong?" I grinned and lifted my eyebrows in reply, though my smile felt just the slightest bit tight. It occurred to me that my employer of the moment, National Geographic Television, had inexplicably chosen to pay me through a temp agency.

Then I climbed up onto the rim of the tank, where a former curator and piranha buff named David Schleser was waiting to give me a primer. "It's not pir-anna, it's pir-*an*-yaa," he said. The name comes from an Amazonian Tupi patois and means simply "toothed fish." Also known in some parts of its range as *capaburro*. Schleser waited till we were both about neck-deep in the piranha tank, and then he added "It means donkey castrator."

The Devil of Evil Wild Nature

There are at least two running themes in the mythology of the killer piranha. The first is that swarming, blood-maddened hordes of these little fish will strip to the bone any creature dumb enough to wade into the South American lakes and rivers where they live. At least in the English-speaking world, the piranha owes its savage reputation largely to Teddy Roosevelt, bestriding the untamed planet in his knee boots and pith helmet.

In 1913, after an unsuccessful attempt to reenter the White House, and in search of new worlds to conquer, Roosevelt undertook an expedition to Brazil. He wrote with gusto about "the devil of evil wild nature in the tropics," and he found the perfect incarnation of it in the piranha. On the second day of his trip, when he had hardly gotten his feet wet, he sent a letter to the newspapers back home. "Piranhas," he wrote, "are the most ferocious fish in the world . . . They will snap a finger off a hand incautiously trailed in the water . . . They will rend and devour alive any wounded man or beast." He recounted a fellow traveler's tale of a man who went out riding on a mule—and the mule returned to camp alone. Next day, the man's companions found his skeleton in the water, "his clothes uninjured but every particle of flesh stripped from his bones." In the letter and Roosevelt's subsequent book, *Through the Brazilian Wilderness*, piranhas captured the fevered imagination of his eager readers.

The flesh-rending piranha thus entered the pantheon of the immortal villains, where it has remained on up through such dismal films as *Piranha II* (in which piranhas somehow get crossed with flying fish and fly through the air with bared fangs, saying, *graak-graak-graak-graak*) and the computer game Tomb Raider III, in which the shapely heroine, Lara Croft, sometimes dies in a pool of seething piranhas.

The second running theme in the piranha mythology is castration, with radical mastectomy not far behind. The earliest known account of piranhas, by the Portuguese explorer Gabriel Soares de Sousa from his 1587 expedition in Brazil,

noted that Indians feared them because "they attack the geni-
tals and cut them off." Likewise an eighteenth-century American naturalist in Guyana reported that "the breasts of women,
and the privities of men swimming in the rivers" are frequently amputated. (He also noted how odd it was to see ducks
stumping around without any feet.) In one highly improbable
case reported in 1920, the alleged victim was a 10-year-old boy
standing knee-deep in a stream who had his penis "amputated
at the base by a piranha that jumped from the water and in a
single slash removed the member."

This is a big reputation to live up to, even for a very bad fish.
That day at the Dallas Aquarium, Schleser told me that of the 40
or so known piranha species only 3, all of one genus, present any
real danger.

"And these?" I asked.

"These are one of the dangerous ones," he said. They were
the notorious red-bellied piranhas, genus *Pygocentrus*, and our
toes had hardly touched the surface of the water when they fled
to the far end of the tank and cowered. Our pasty white legs went
unmaimed, and the seven-year-olds on the other side of the glass
glumly moved on to other tanks, with other terrible possibilities.

But what red-bellied piranhas will do in the confines of an
aquarium isn't necessarily what they will do in the wild. "I consider piranhas as spiteful creatures," wrote Dr. George S. Myers,
author of *The Piranha Book*, in 1972. "They *must* be fed with flesh;
it can be seen in their eyes, in their movements, and the stupid
expressions on their faces."

Getting Skeletonized

With that cheering thought in mind, I departed one day from
Iquitos, Peru, on the *Esperanza*, a riverboat of the kind known

Reality Television?

I was working for National Geographic on a show about piranhas at a time when everyone there was under orders to cut costs. The days were sadly past when the Society could sustain an explorer trekking for years through the remotest parts of China equipped with portable bath and full dinner service (not to mention the rumor of a traveling catamite). At other companies, the brief fits of penny-pinching typically translate into paperclip quotas and the like. And at National Geographic? The assistant producer, an eager beginner, had bought me a red bathing suit so I would be clearly visible on camera when I climbed into the tank of piranhas at feeding time. ("Do piranhas like red?" she'd inquired thoughtfully.) And when I climbed back out of the tank after the shoot, still dripping but unmolested, she was waiting for me with her clipboard. "I need the bathing suit back," she said, pointing distastefully with her pen. It turned out that she'd held onto the price tag, so she could return it as defective merchandise.

On the other hand, traveling with a television crew could also be a delight, mainly for the entertaining cast of characters. On one show about prairie dogs, the soundman was a gentle soul, gray-haired, a little paunchy, with a fondness for music by Tom Waits and 3 Mustaphas 3. He made a point of being almost gushingly thankful to everyone we worked with: "You were so wonderful to let us into your home. Thank you so much. You've been so patient with us."

It was instructive watching him, because he never said anything bad about anyone, whereas most of us never said anything good. "I know that many people I have loved and admired have also been devious, despicable people," he admitted, when we questioned this behavior. But one day he'd stepped outside himself and didn't like what he saw when he actually expressed these feelings. Nor did he like leaving behind that wake of poisoned feelings. The rest of us took this lesson deeply to heart.

And next day, we made it our special mission to get him to say something rotten before the end of the shoot. Lord knows, he had plenty of material to work with just within our own little group. (The cameraman, for instance, liked to look up soulfully at pretty waitresses and present his order as if "chicken piccata" were a love note.) We were also filming colorful characters, including a welder whose life had been transformed one night by a dream about a big yellow truck with a green hose for sucking prairie dogs out of their burrows. (Now he had just such a truck and was charging $1,500 a day for the service.) There was also a biologist who claimed to be able to interpret prairie dog talk, and a former military sniper whose hobby was vaporizing prairie dogs at 1,000 meters.

But the soundman remained unfailingly kind. We finally had to resort to alcohol. On the last night of the shoot, the producer, no tightwad, took us to the best restaurant in town and ordered some terrific wine. Somewhere around the fourth bottle, the soundman suddenly started talking about a neighbor, a famous novelist back home in Cape Cod. First he likened her to a certain bulbous mushroom, red with white speckles. No, no, wait, he said, she was actually like the characters in a certain Japanese science fiction film who ingest mushrooms in some alien world, become giddy, and then find that their heads have flattened out and sprouted gills. On a roll now, the soundman added that this fishlike humanoid was married to a writer who once appalled a New England audience by reading his account of masturbating a cat with a Q-tip.

There was a brief shocked silence. It was if the Dalai Lama had confessed to a fondness for Paris Hilton videos. Then everyone cheered. The producer ordered two more bottles of good red wine, and we drank a toast to the rarest thing in the wonderful world of television—reality.

locally as a *pamacari*, like a square-ended canoe, 12 feet across at the gunwales and 60 feet long. It had a cabin with bunks for eight, and was equipped for a weeklong fish-collecting expedition led by Schleser. Our party consisted of the sort of people who could identify lunch not merely as catfish, but as *Brachy-platystoma filamentosum*. They passed the travel time by trading fish chow recipes (combine beef heart, peas, spinach, and live goldfish in the blender) and by guessing what the pH might be in this stretch of water.

We were soon up to our waists and even our necks in the muddy Amazon and its tributaries, pulling seines and flailing dip nets. At the end of each haul, we knelt around the nets on the shore to inspect our glittering catch. "What have we got here?" Schleser said. "River dogs and raphiodons?" Many of the fish we caught had the evidence of piranha attack on their hindquarters—but not the sort of bloody stumps or open wounds you might expect. The damage consisted mainly of small divots the shape of toenail parings, which piranhas had neatly clipped out of the trailing edge of tail fins. "So we know they're in here," Schleser said.

In fact, the river seemed to have no shortage of threatening species. There was a catfish shaped like a miniature blue whale, known locally as the *carnero*, which can bore Swiss-cheese holes into fish trapped in nets or, on rare occasions, scoop melon-ball chunks out of people swimming. There were anacondas and electric eels. There was also a stingray up to 5 feet across, which can drive the poisonous barb on its tail into a person's leg. "Shuffle your feet on the bottom," Schleser advised. "They'll go away if you bump them."

In one haul, we collected a slithery, eel-like little catfish with no pectoral fins or dorsal spines. Schleser identified it as the notorious candiru, a parasite, which typically swims into the gills of other fish, latches on, and pierces a gill artery to feed on the blood. The candiru is also reputed to enter the genital region of anybody dumb enough not merely to swim in the Amazon, but to urinate while doing so. When I held the candiru in my hand, it dug in with two little spines projecting down from either side of its mouth. The candiru is said to use these spines like pitons for climbing up the urethra.

Piranhas hardly seemed to worry anyone. In almost every village along the river, people knew someone who had been bitten by them, and in almost every village, small children flung themselves gleefully from the riverbank to swim. Piranhas only seemed to bite when they were flopping around in the bottom of a fishing canoe. "A mouse will bite, too, if you pick it up by the tail," Schleser said.

One day, we stopped on the bank of the Rio Napo, a tributary of the Amazon, and threw out a couple of fishing lines baited with liver. Red-bellied piranhas soon started biting. They didn't swarm. They didn't make the water boil. They behaved, in fact, like any other fish, tugging at the line, stealing bait and sometimes getting caught. We kept some of them for lunch and released a few others, keeping our fingers clear of their toothy, eager mouths. It was getting toward midday and the equatorial sun was blazing down. We let another piranha go, and on a whim I waded into the river after it to cool off.

"We're chumming piranhas with raw meat and you're going swimming?" one skeptical member of the party inquired. He was standing on a log across a little inlet. A moment later, the log gave way and he sank straight down into the water, as if he were riding an elevator and the piranhas down in the lobby had just pushed the call button. He was still alive when he came back to the surface, merely a little refreshed. The rest of the party soon joined us in the river, passing around soap and shampoo, and no one was skeletonized, except the piranhas themselves, which we ate for lunch.

Frightened Cows

The truth seemed to be that we eat piranhas far more often than they eat us. But they will sometimes return the favor, according to a paper with the colorful title, "Scavenging on human corpses as a source for stories about man-eating piranhas." Together with a medical colleague, ichthyologist Ivan Sazima of the University of Campinas in São Paolo investigated two drownings and one heart-attack death in the waters of the Pantanal region of western Brazil. One body was recovered after four days and was almost completely skeletonized (an important word in the piranha argot), another came out after 20 hours with the appendages but not the torso reduced to bone, and the third came out after just a few hours having suffered what amounts to a severe complexion problem. Red-bellied piranhas were the main scavengers in all three cases, according to Sazima. He suggested that cases of postmortem scavenging like these probably gave rise to the idea that piranhas will attack and kill humans swimming in infested waters. But Sazima could not locate a single authenticated case in which such a fatal attack had actually occurred.

So what are the piranhas eating, if not us? To find out, Sazima and a colleague at the University of Mato Grosso spent more than 300 hours snorkeling in the clear-water ponds and creeks of the Pantanal, observing the behavior of the fish in their natural setting. The evidence was that most piranhas, like most men, lead lives of quiet desperation. Instead of swarming over their victims in a tumult of flashing teeth, the piranhas mostly lurked and stalked and even disguised themselves as other species to snatch their food on the sly. They tended to bite their victims on the tail and then run away. Only one of the three species observed was a genuine killer: the red-bellied piranha. But they also tended to be furtive, even though they traveled in groups of 20 or 30 fish. They often hunted by hiding in the vegetation and dashing out to blindside small fish, which they swallowed in a bite or two.

The other piranhas in Sazima's study fed mainly on scales and fins, not meat. In the 1970s, a researcher who was still attached to the fierce image of the piranha conjectured that attacking the tail fins was the equivalent of wolves severing the hamstring of a deer, to cripple the prey for an easier kill. But the terrible truth, according to Sazima and other recent researchers, is that most piranhas seldom actually proceed to the kill. A piece of tail fin, or a section of scales raked off like roof shingles, is all they're really after.

Leo Nico, a researcher at the U.S. Geological Survey's biological research station in Florida, writes that fish scales and fins can be anywhere from 34 to 85 percent protein. They may also provide other vital nutrients in the calcium- and phosphorous-poor blackwater rivers of South America. "Why take fins or scales when whole fish are more nutritious?" Nico asks. The answer appears to be that it's easier to steal a mouthful of scales or fins than to kill a whole fish. Moreover, both scales and fins grow back in a few weeks. They are a renewable resource. Thus, Nico writes, "Piranhas resemble herbivores that continually exploit their food source without eliminating it." Think of cows cropping grass in a pasture.

Or rather, think of frightened cows. In Sazima's study, one fin-eating piranha species preferred to sneak up on its victims from behind and below. Then it would strike at the tail or anal

fin, clipping off a piece with an audible snap and twisting its body
to tear it away. The young of another piranha species sometimes
joined schools of similar-looking cichlids and lollygagged along
pretending to be just one of the gang. Every now and then, they'd
snap a bite out of the tail fins of the real cichlids and then swim
along again looking innocent.

This sort of sneaky behavior is the reason the tail fins of fish
in piranha habitat commonly exhibit what one researcher calls
"small lunate regions of loss," a lovely phrase, which might also
describe the human heart in middle life. And not too surprisingly, the fish do not much like it. Losing chunks out of their
tail fins can affect their swimming ability and thus their feeding
and mating success. In Sazima's study, many prey fish seemed
to know exactly what the piranhas were up to. When piranhas
approached, one group of cichlids circled their wagons, forming
a ring with their tails toward the center, and then moved together
to the shelter of vegetation.

Other species defend themselves ferociously. Kirk Winemiller, an ichthyologist at Texas A&M University, describes "the
pointed conical teeth and highly protrusible jaw apparatus" of
one cichlid, and its knack for turning on the piranhas with its
mouth open and membranes flared. This is apparently sufficient
to send timid piranhas bolting for shelter. The piranhas adapt by
trying to catch their prey unawares, attacking them when feeding, fighting, or courting distracts them. They like to feed on a
slow-swimming cichlid species that noses around aquatic plants
picking off minute invertebrates. This feeding behavior frequently leaves the cichlid in the ludicrous position of having its
face buried in floating vegetation and its hind end hanging out
in the water below, like a neon sign saying, "Bite here." But this
cichlid has evolved eyespots on the tail, which also has the same

coloration as the head. It's like a neon sign saying, "I might bite back." The resulting "startle effect" may be enough, Winemiller writes, to confuse a piranha rushing in for a sneak attack.

Because so many piranhas seem to prefer scales and fins to red meat, researchers now commonly describe them as behaving more like parasites than predators. Sazima writes that "they are unique among predatory fishes since they are able to give an individual prey a disagreeable experience" not just once in a lifetime, but repeatedly, like an overbearing boss, or a nagging spouse. But Sazima also watched one piranha species approach another and politely pick the fish lice from its flanks, taking nothing else. Piranha feeding behavior may have started with this sort of social grooming, according to one theory, and evolved first to fin- and scale-nipping and then to actual rending of flesh. A few piranha species do not even eat scales or fins. They feed mainly on seeds, nuts, fruits, and even flowers. This is an image for our time: The New Age Piranha. The piranha as crypto-vegan.

But the object of my journey wasn't to turn the piranha into a petting-zoo animal. In my heart, I was still hoping to discover some remnant of "the devil of evil wild nature." Venezuela sounded promising.

Yaquelin Was Here

Los llanos—literally "the plains"—is a region of flat grassland and dry forest in the interior of the country, veined and capillaried with hundreds of little streams running down from the Andes Mountains to the Río Apure, and thence to the Orinoco. In the rainy season, the rivers flood the land and when the water subsides, it leaves ponds and lakes almost everywhere. The *llanos* is cowboy country now, so the watering holes tend to be fringed with cream-colored Brahman cattle and sometimes water buf-

falo. They're also fringed with some of the most spectacular concentrations of water birds in the Western Hemisphere. One day
at sunset, as I made the long drive to the town of San Fernando
de Apure, on the Río Apure, I saw 50 or so white egrets roosting
together in a tree beside one such pond, covering it like Christmas ornaments. An ibis flew past, red as a rose, sleek as a woman
in a tight little party dress, and joined a dozen others just like it
standing around in a gaggle.

The birds thrive on fish that get left behind in every lagoon
and watering hole, red-bellied piranhas almost always among
them. The concentration of the fish, and the intensity of feeding
by both birds and piranhas, increase as the watering holes dwindle in the dry season. Under these circumstances, I'd been told,
piranhas sometimes act like piranhas.

On the road that day, an odd little scheme came to mind. It
had to do with the two dead iguanas I passed, and the boa carcass as thick as my thigh. I stopped at a sort of general store to
buy some rope and, imprudently, I began to explain my brilliant
idea in badly broken Spanish. It would have been a difficult idea
to convey even in English, or to imbue with the faintest shred of
reason. In Spanish, it took me a long time just to figure out how
to say "road kill." But after a while, I was able to get across the
idea that I wanted to pick up a dead animal, tie its stinking corpse
to the hood of my little Fiat, drive it a couple of hundred kilometers, and then fling it into a river full of piranhas to see just
how long it really took them to eat it. I might as well have been
saying I wanted to pick up a dead animal and take it back to my
hotel room because, you know, it does get lonely on the road. The
shopkeeper declined to sell me any rope.

My hotel room that night was just the sort of place to shack up
with road kill. It was like a slightly luxurious jail cell. That is, there
was a seat on the toilet, and a shower, and both of them leaked and

mingled their waters on the bathroom floor. The names "Carmen y Pedro" and "Edgar y Yaquelin" were carved into the rusty bedstead. There was a light on the concrete wall with one bare bulb.

I arrived in San Fernando de Apure next morning with only my own sorry carcass—and the name of a contact at FONIAAP, Venezuela's national fisheries agency. The contact was apparently accustomed to northern visitors with odd ideas about his continent. He politely ignored my assertion that I knew the location of a dead dog in the neighborhood (I think it was dead, anyway), and set me up with a capable biological technician named Orlando Camacaro. Camacaro and I quickly worked out a protocol for sampling local piranha populations. Then we went out and stocked an ice cooler with freshly slaughtered beef shank, bone-in. I suggested that a couple of 3-pound chickens might also be a good idea. The shop around the corner was selling them frozen and wrapped in plastic, but I explained to Camacaro that I wanted a *pollo completo*, feathers and all. I didn't know the word for feathers. So I ran my hands over my ribs, tapping my fingers up and down and thinking to myself that, at times, we *norteamericanos* must make a very unusual impression.

"Leeve?" Camacaro asked, by which it dawned on me that he meant "live."

I ran a finger across my throat.

We soon had two 3-pound chickens in the back of the Fiat, complete with feathers and feet, and we set off in search of piranhas. Our first stop was a backwater of the Río Apure, where two fishermen were gutting their catch at a roadside stand. They were mostly selling catfish, but they also had stacks of fresh piranha for sale. Powdered piranha sells well, too, said one of the fishermen. Though this might seem paradoxical for an animal sometimes known as the "donkey castrator," it has a reputation as an aphrodisiac. "It's good for the brain," said the fisherman, grinning. "But it's good for the rest of the body, too."

Down by the river, Camacaro tied a chunk of beef on a line. The plan was to put it in the water for a minute at a time and see if we got any reaction. He tossed the meat 5 feet from shore and I asked, "What time is it?" But before I could get out the words, Camacaro had already jerked the line back to shore with a half dozen piranhas attached. They landed at my feet, their orange bellies glistening, their tails frantically slapping. They flipped and wriggled their way back down to the water. They were small fish, just 3 or 4 inches long, but I was impressed.

Then one of the fisherman threw in a dead fish for our amusement, and the water seemed to explode before us. As the carcass rolled over and over, piranhas with their teeth sunk in its flanks did somersaults. Sometimes a piranha shot across the surface with a sound like a whip lashing through the air. The secret of this extraordinary display, said Camacaro, was that the fishermen constantly toss fish guts into the water here, keeping unnatural concentrations of piranhas in the area more or less permanently.

A couple of cormorants were swimming in the same water, just a hundred yards away. But even at that short distance from the constant chumming of fish guts, the natural order seemed to be restored. One of the birds vanished beneath the surface—and came up after a moment with a gold flash of piranha held crosswise in its bill. It mouthed its catch for a while, turning it as if for a better position. Then in a smooth, well-practiced move, it flipped the piranha around and swallowed it whole, headfirst. I watched for a minute, to see if the piranha, with its ferocious teeth, was going to come rotoring right out the cormorant's other end, like some sort of tunnel drill. But the bird took off and flew away, apparently content.

It went like this for much of the rest of the day, as Camacaro and I traveled from one watering hole to another. Often, we got no response to our bait, or the piranhas merely picked at it, idly. A few times the piranhas hit the meat immediately and sent back only the polished bone, with the marrow hollowed out. But these occasional displays of ferocity seemed to be a product mainly of unusual circumstances.

We stopped, for instance, at a ranch named Hato La Guanota, which keeps 4,000 water buffalo on pastures that are some-times flooded by the Río Apure. The manager, Hector Scannone, explained that water buffalo, being thick-skinned and sensitive to heat, like to spend much of the day wallowing in water holes. This is just too tempting for the piranhas to resist. They dart in and bite the teats off the udders of the heifers and the testicles off the bulls. This puts a crimp in Scannone's business of mak-ing mozzarella cheese. When the ranch first started grazing water buffalo in the 1970s, Scannone said, 80 percent of the herd suf-fered piranha damage. Among other desperate measures, he tried to dynamite the piranhas out of the water holes. Now, he simply drains the ponds temporarily and removes the fish when the piranha population becomes too large. He still has to get rid of about 30 damaged animals a year. But like the fish-gutting areas, this was a man-made opportunity for the piranhas.

I came across just two natural circumstances in which pira-nhas display their stereotypical swarming behavior: In ponds that have dwindled to the point of starvation and in the waters beneath bird rookeries. The huge bird populations of the *llanos* tend to nest in trees over the water, according to Don Taphorn, an ichthyologist at UNELLEZ, one of Venezuela's national uni-versities. When the birds bring back food to regurgitate for their young, it invariably gets spilled into the water, along with their droppings.

"This chums in the piranhas because it's free lunch, break-fast, and dinner," Taphorn said. "So the rookeries are packed solid with piranhas just waiting for something to fall." Taphorn once saw an anhinga fledgling, a big bird like a cormorant, fall into the water at a rookery.

"He was trying to get back to the tree and the piranhas wouldn't let him," Taphorn said. "You could see all this boil-ing water, and at the same time, the caimans were coming off the shore. It was like something out of a Tarzan movie. But it doesn't happen often." In fact, it was the only time in 22 years that Taphorn has seen piranhas attack a living animal. Though he has spent much of his career neck-deep in the waterways of the *llanos* pulling fishnets, he has never been bitten himself.

Bad News in Boquerones

Camacaro and I were too late in the season for the bird rookeri-es. But just about the time I was giving up on the idea of pira-nhas attacking much of anything other than fish tails, we arrived at the fishing village of Boquerones. We turned off the road at a bridge and went down a steep bank of dust as fine as powder. A wide, ramshackle concrete stairway under the bridge ran down to the water's edge. The local fishermen land their long metal riverboats here, and they also inadvertently chum the water with fish guts, attracting a large piranha population. We decided to try out a chicken carcass, as a fisherman sat nearby patiently mending his nets. The piranha hit the carcass the instant it touched the water. After one minute and 20 seconds, they'd eaten everything except the spine and leg bones. At two minutes, the water was quiet again, and there was nothing left but three wing feathers drifting on the surface. The fisherman never even looked up.

By now, it was pretty clear how the mythology of the piranha had evolved. The locals, whether they were Amazonian tribesmen or early settlers, have always had places where they gutted their fish. And they have always been blasé about the piranhas that assembled there, much as people in New England, where I live, are blasé about the ferocity of bluefish. But an explorer passing through could hardly help being impressed. And it would have been entirely natural to attribute the feeding behavior at these hot spots to all piranhas, and to all South American waterways. All it needed for the piranha to become legend was the occasional nasty accident . . .

Camacaro and I went over to the little blue grocery on the riverbank, where the owner sat in the darkened interior and handed out her merchandise through a metal grate. Her name was Yelitza. Camacaro chatted with her as he sipped a soda, and it came out after a while that, just a month earlier, an accident had happened at the foot of the stairs. Her 6-year-old, Edouardo, and her 11-year-old, Carolina, had been playing on the stairway. Edouardo, who cannot swim, tripped and tumbled into the water, and Carolina went in after him. The piranhas attacked them immediately. A boy on shore had gone in to rescue them both and they pulled one another out in a chain. No one died.

But the two children had just spent three weeks in the hospital, and Yelitza showed us their injuries. Edouardo, who'd been wearing blue jeans the day of the accident, had a long scar across the back of his right calf. Carolina had been wearing shorts, and the piranhas had opened long half-inch-wide wounds across the tops of both of her thighs. Then, one piranha biting in the same place as another, they frantically deepened the wounds. Carolina had been bitten almost to the bone on one leg and to within a centimeter of her femoral artery on the other, all within a matter of seconds.

What had happened to Edouardo and Carolina was of course

a rare accident. Intellectually, I knew that it was irrational to demonize animals for circumstances we humans have largely created. I also knew that humans do not loom nearly as large in piranha psychology (or piranha stomachs) as piranhas do in ours. They'd rather be off nipping some poor cichlid on the tail.

But several days went by before I ventured back into the water (being careful first that no one was gutting fish in the vicinity). And you can be sure I did not skinny-dip.

Intimate Friends

*T*he habitat was deeply inhospitable—a sheer bluff, knotted and furrowed by subsurface tremors, intermittently flooded, buffeted by winds, burned by the sun. My guide was Cliff Desch, a mild, likable University of Connecticut professor with unruly gray hair winging out over the tops of his ears. We were searching for life on the human body or, more precisely, on the hostile terrain of my own forehead. I took a bobby pin, as instructed, and scraped the crook of it hard across the skin in front of my hairline. Then, like a fisherman emptying his nets, I spread my catch on a glass slide.

The human body, especially the face, is the natural habitat for two species of mites, Desch said, as he placed the slide under a microscope. One species is minutely adapted to the hair follicle. The other ensconces itself in the micro-habitat of the sebaceous gland, less than a millimeter away. Sir Richard Owen, better known for naming another buried life-form, the dinosaur, brought the follicle mite to the attention of the world in the 1840s. He called the genus *Demodex*, meaning "lard worm" (though mites are actually distant relatives of spiders).

Desch peered through the microscope and said, "Oh wow" and then, "Hunh!" It appeared that my forehead was home to only one species of mite. But quickly, before I could become despondent about inadequacies in my personal biodiversity, he added: "You've got the best population I've ever seen."

It occurred to me first that Desch had spent an entire career looking at this sort of thing and second that I had stood under a shower just a few hours earlier, slathering my forehead with soap and blasting it with steaming water. "Look at 'em all," Desch was saying now, unable to suppress his delight. "Holy moley!"

Well, no man is an island. He is an ecosystem, though we studiously pretend otherwise. Our skin—2 square yards (1.7 square meters) of it on the average human body—is a habitat for roughly as many bacteria as there are people in the United States, for fungi and viruses, and on occasion for mosquitoes, fleas, bedbugs and kissing bugs, blackflies and botflies, lice, leeches, ticks, and scabies mites, which tunnel across the backs of an afflicted person's hands like moles burrowing in the front lawn.

In the developed world we like to think we have tubbed and scrubbed ourselves free of any overly personal connection to the natural world. Even mosquitoes stay mainly on the other side of our window screens. But this is a delusion, as follicle mites, which live on almost everyone, abundantly demonstrate.

I stepped up to the microscope, and they came into focus, lying crisscross like sticks of wood. The adult mites were about a hundredth of an inch (0.25 millimeter) long. Their stumpy little legs wriggled and twitched as in a dream. They had tiny claws and needle-like mouthparts for consuming skin cells. Here and there were eggs shaped like arrowheads and juveniles with angled-back scutes on their underbellies, like fish scales, the better to anchor themselves in my skin. Desch eyed my forehead as if it were the Grand Banks in high season and said, "I think it's great." I smiled wanly.

The Erotic Flea

Once upon a time we were all far more at home, though not necessarily any happier, with the idea of being infested. A fifteenth-century courtier once discreetly picked a louse off King Louis XI of France, and the king graciously remarked that lice remind even royalty that they are human. (Next day an imitator pretended to find a flea on the king, who was by then perhaps tired of being human. "What!" he snapped. "Do you take me for a dog, that I should be running with fleas? Get out of my sight!")

For almost all our history as a species, being infested was an inescapable fact of life, and our forebears achieved an intimacy with nature that we can scarcely imagine. European lovers of the seventeenth century sometimes wrote seduction poems about a girlfriend's fleas. John Donne once petulantly complained that a promiscuous flea, having bitten boy and girl alike, "swells with one blood made of two / And this alas is more than we would do."

A few gallant Frenchmen actually plucked a flea from their lady love and kept it as a pet in a tiny gold cage at the neck, where it could feed daily on their own blood. In Siberia, according to one story, an explorer was disconcerted to find that young women visiting his hut tossed lice at him; it turned out to be their way of expressing amorous intentions.

Clearly, this would not be a successful dating strategy today; for one thing, the human flea itself has almost vanished from modern homes. The hardier cat flea has replaced it, but only partly. Body lice, too, are far more scarce; they lay their eggs in our clothing, an elegant adaptation to human hairlessness, but have thus fallen victim to that environmental cataclysm, the rinse cycle.

The more remote our ectoparasites have become, the more horrifying they seem to be. Moreover, science has made this horror seem rational by demonstrating over the past century that several of our ectoparasites are the most dangerous animals on Earth. The diseases they carry have killed us by the hundreds of millions—fleas with bubonic plague, body lice with epidemic typhus, mosquitoes with yellow fever and malaria. They vex and panic us even in the most modernized countries with maladies like encephalitis, transmitted by mosquitoes and ticks, and tick-borne Lyme disease.

We go to sleep at night aware that our very pillows are home to thousands of dust mites—which, as it happens, help keep our homes clean by busily consuming the tens of millions of skin cells we shed each day. But the mites also cause asthma in some people, and when it comes to the beasts that live on and around our bodies, we tend to focus on the negative.

So it takes an almost unnatural objectivity to suggest that our ectoparasites can also be fascinating. Like any species colonizing difficult terrain, they have adapted ingeniously to our flesh. They use sophisticated chemosensors to find us; saws and scalpels to penetrate our skin; siphons and a small pharmaceutical ware-house, including anesthetics and anticoagulants, to steal a blood meal and get away undetected. If we can suspend for a moment the uneasy awareness that all this evolution is geared to extract-ing our blood, and if we can forget that our parasites mostly use this blood to produce the eggs for their future pestiferous gen-erations, then it is possible to regard them with awe.

Long-Distance Parasites

They are capable of extraordinary subterfuge. For example, the adult botfly (*Dermatobia hominis*) of Central and South America manages to parasitize us quite gruesomely without ever actually making physical contact. To avoid being swatted by some balky human or other host, she captures an insect, a mosquito for example, glues her eggs to her prisoner's abdomen, and then sets it free.

The mosquito ignores the eggs (as will we for a moment) and goes off to employ subterfuges of her own. Many mosquitoes feed at night, for obvious reasons ("Consider the outcome if you were to approach an elephant with a syringe," one entomologist says). But this mosquito is a day feeder, finding a victim with her eyes and with sensors attuned to carbon dioxide, warmth, lactic acid, and other bodily emanations.

Having deftly touched down, the mosquito slices her way into the fine web of blood vessels in the skin. The damaged vessels instantly attempt to plug their leaks with aggregating platelets in the blood. But host and parasite have evolved together, with all the one-upmanship of any arms race. So the mosquito is equipped with a powerful enzyme in her saliva to disable the platelets. The more saliva she pours down one tube in her proboscis, the faster she can suck up blood through another. Humans in turn have an immune response to the saliva, which alerts us with itching and swelling, but only after about a minute. We swat ploddingly—and are likely to kill only the slowest feeders. Thus we do our bit for natural selection, helping ensure that future generations come only from mosquitoes that are quick enough to get away with our blood in a minute or less.

But the coevolutionary arms race on the human ecosystem is even more disheartening than all this might suggest. The mosquito may leave behind other gifts, along with her saliva. After having been driven out in the mid-twentieth century, malaria and dengue fever have lately begun to reappear in the United States and other developed nations. Insect-borne diseases are on the increase worldwide: air travel and economic globalization have brought distant nations together. But control measures tend to get poorly funded, and in some cases insect species and pathogens have developed resistance to standard treatments.

In the New World tropics the insects may arrive bearing not just agents of disease but at least one other gift: Let's say we get bitten by the mosquito that was briefly held prisoner a few days earlier by a botfly. As the mosquito feeds, our own body heat triggers the botfly eggs glued to her abdomen to hatch. A botfly larva promptly crawls into the fresh bite wound, where it matures with time into the ripest sort of traveler's horror story.

The larva has a segmented, yellow-brown Michelin-man body, belted with rows of raked-back spines for lodging itself mouth-first in the skin. It also anchors itself with two tusklike hooks sticking out from the mouth. Its tail is a breathing tube, which can lift up, periscope-like, just above the surface at the point of entry. As it develops, the larva wriggles visibly and painfully under the skin. Removing the botfly is relatively simple (one remedy involves applying bacon to the breathing hole, so the botfly has to burrow up through it for air). But a Harvard biology student, curious about his own potential as an ecosystem, once nurtured a botfly in his flesh for six weeks. Finally a 1-inch-long (2.5-centimeter-long) botfly larva, ready to move on to its pupal stage, started to emerge from his scalp as he sat in the bleachers during a Red Sox–Yankees game at Fenway Park. The Sox lost, and despite the biologist's heroic efforts to protect it, the botfly died.

Parasitic Hysteria

But the beasts that live on our bodies are by no means all bad. A normal population of bacteria on the skin, for example, may actually benefit us by preventing infectious bacteria from gaining a beachhead. But if you tell people that a normal population can mean a hundred bacteria per square inch in the barren habitat of the shoulder blades (or millions in the sweltering armpit), they are liable to scrub themselves raw. In the extreme disorder called delusory parasitosis, victims can imagine they are under assault by invisible bugs that spill out of electric sockets, crawl from holes in concrete, and drop down from ceiling tiles. To stop the constant itching, they scratch themselves bloody. They bathe in gasoline and inundate their homes with pesticides. But the bugs keep coming. Such cases have sometimes ended in suicide and once in the murder of a doctor who tried to get his patient to see a psychiatrist.

When real infestations occur, even sensible people often behave irrationally. In the course of their recent evolution, for instance, head lice seem to have developed resistance to most conventional treatments. Distraught families of infested school-children frequently resort to home remedies. A few years ago in Oklahoma a man applied a highly toxic cleaning solution to a six-year-old's scalp, causing cardiac arrest and permanent brain damage.

So it's important to realize that we aren't under assault or, rather, that the assault is limited and controllable. We possess the ultimate weapon, which is human intelligence—or, anyway, the opposable thumb. In New York City and Boston, profes-sional nitpickers now charge up to $100 an hour for the most venerable treatment for head lice: removing the eggs, or nits, by hand, having first drowned them in a shampoo of olive oil. It is a very old idea of quality time. "It gives you a lot of bonding when you nitpick," says Mary Ward, a Boston nitpicker. "You know these people."

Our ancestors would regard our otherwise unpestilential lives with dumbfounded envy: We don't spend our days itching and fidgeting; we know which diseases our parasites carry and how to avoid them; and at least in the more temperate corners of the planet, we don't generally suffer from nightmarish stuff like botflies. Despite panicky early concerns, we know that our ectoparasites do not transmit the AIDS virus.

This peace to which we have grown accustomed may be more fragile than we care to acknowledge: In 1999, mosquito-borne West Nile virus was introduced into the United States from Africa and is now permanently established. In 2007, a traveler arrived in Northern Italy from India suffering from Chikungunya, a tropical disease whose name in the East African Makonde language means "that which bends," for the arthritic posture it can produce in its victims. Tiger mosquitoes, an introduced species, spread the disease to 292 people in the city of Ravenna, and 1 person died. Still, it's much worse in the developing world. On the island of Reunion the year before, Chikungunya killed 236 people. Mosquito-borne dengue fever kills an estimated 20,000 people a year in tropical developing countries, and the annual death toll from malaria is well over a million. In the developed world, says Duane J. Gubler, an infectious disease expert at the University of Hawaii, "we have better hygiene, houses with screen windows, air-conditioning. Television has made us reclusive, at home at the time when we are at greatest risk of being bitten by mosquitoes."

We are spared by being couch potatoes, each of us a lonely and underpopulated habitat, perched before our television sets, with only our resident bacteria and those low-key hangers-on, the follicle mites, for company.

I thought about all this as I looked through the microscope in Cliff Desch's laboratory. I also thought, as so many of us do in moments of aesthetic and personal doubt, about Martha Stewart, who has written "I have always been inspired by nature." I asked

Desch what sort of inspiring things the follicle mites might be doing on *her* forehead and by extension on riffraff like me.

These mites, he said, aren't much good at crawling to new territory. But they spread from person to person when we nuzzle, and because a population thrives in the area around the nipples, they also pass to newborns as naturally as mother's milk.

An immigrant mite makes itself at home on a fresh face almost instantly, crawling mouth first into the nearest follicle, with its back to the hair shaft and its stumpy legs to the follicle wall. Since it has no reverse gear, Desch said, it may never come out again. Embedded upside down in our skin, it feeds by using those needlelike mouthparts to puncture epithelial cells and suck up the spilled fluids—with no apparent harm to us. It filters out solids even as small as the mitochondria of the cell, a feat Desch characterized as "near-perfect pre-oral digestion." The mite's digestive process yields so little waste that it doesn't even have an excretory opening. It need never get up to go to the bathroom. The follicle mite is, in truth, a couch potato's couch potato.

"And to reproduce?" I asked Desch, with some trepidation, thinking that a mite must get lonely tucked away somewhere out on the vast, windswept expanse of the forehead. The nearest neighboring mite population centers, around the wings of the nose and in the eyelashes, are as distant as oceanic islands.

The female, Desch said, may produce a first generation asexually, by parthenogenesis—that is, virgin birth. Then she mates with her sons to produce the next generation, up to a maximum population of about 10 mites per follicle. ("Oedipus should have plucked out his eyelashes and left his eyes alone," I muttered.) All this passes utterly unnoticed, "the extreme," one biologist remarks, "of an exquisite adaptation in which each of us is infested right now, but asymptomatically." Some researchers theorize that follicle mites may even benefit us in ways we do not yet understand. In any case, there is nothing, from soaps to systemic medicines, that we can do about it.

I left Desch's lab thinking that follicle mites are precisely the ectoparasite we deserve—and that we are lucky to have them, riding on our foreheads, a living reminder that our flesh is merely a part of the natural world.

Back home I offered to write my wife an ode to her follicle mites. She handed me a washrag for my forehead and suggested curtly that I keep my infestations to myself. But I knew that in the nature of life on the human habitat, it was already way too late for that.

In the Realm of Virtual Reality

In the courtyard of a monastery somewhere in central Bhutan, cockerels strut across the lattice of flagstones. A dozen temple dogs doze fitfully in the sun. A novice late for prayers scurries, sandals in hand, and vanishes into a doorway. Inside, it's cold and dark. The only light oozes down like buckwheat honey through a narrow atrium, dimly illuminating the weird figures painted on the heavy timbers, and the monstrous antlers of a supposedly extinct deer, the shou, lashed to a balcony railing.

From a black recess up ahead comes the low, sonorous sound of chanting, rising steadily in pitch and fervor. A horn made from a human femur goes *oh-woe-oh-oh-oh*. Cymbals stutter. A pair of short metallic horns put up a high-pitched, reedy sound. A drum beats. Our guide leads us into a narrow hall and then, after we remove our shoes, into the prayer room, where monks in maroon robes sit around the periphery and a smudge of incense drifts up from a censer. As our eyes adjust, strange animal shapes form vaguely out of the darkness: the dust-cloyed head of a tiger hanging on a wall, a huge, primitive-looking fish with long bony scales, a string of human hands. And off to one side of the altar, the thing we have come halfway round the world to see, a hanging figure with a white veil draped over its head.

Voice of Reason

Legend says a holy man brought Buddhism to the Himalayan kingdom of Bhutan 1,200 years ago, flying in on the back of a tigress. Today, you get to Paro International Airport in a Druk Air 72-seat jet. It's reputed to be the most technically difficult landing in the world. "While flying in," the pilot announces on the initial descent, "you may find yourself coming closer to the terrain than is usual in a jetliner. Much closer. This is normal. Don't be alarmed." Most pilots crash several times before learning how to thread their way down the corridor of valleys to the airstrip. Fortunately, they make their mistakes on a flight simulator before attempting an actual approach. Bhutan has always been a place where the virtual and the real happily coexist.

Having opened itself to the outside world just 50 years ago, the "Land of the Thunder Dragon" is also a place where tradition still shapes everyday life. In the terraced rice paddies of the Paro Valley, families thresh rice by hand, the sheaves swinging overhead sending up plumes of dust, then down, *swot*, on a rock, over and over, until all the dry kernels of rice break loose and rain down in heaps. Chili peppers, another great national food, are spread out in bright red carpets on rooftops and hillsides to dry in the sun. Only about 700,000 people live in Bhutan, most of them in the fertile valleys. Uphill, beyond the last timbered farmhouse, the forest still covers 70 percent of a country the size of Switzerland (only more mountainous). Tigers, leopards, and bears still wander there. So, too, according to legend, does the migoi, which is what the Bhutanese call the yeti.

Migoi literally means "wild man," and the idea that there might be an undiscovered primate, a hairy quasi-human biped, still living in Bhutan's uncharted mountains is, according to some members of our expedition, the stuff of great adventure. And to some of us it is utter bunk. We agree, at least, that seeking the migoi is a way to get beneath the surface of this intriguing culture, which is why a British-American television partnership has sent us here. The group includes an Oxford-trained evolutionary biologist, a primatologist who has spent years working with monkeys in West Africa, and a British technical wiz who will keep our gear in working order. I've been hired to be the on-camera spoilsport or, as I prefer to see it, voice of reason. We have come equipped with camera traps, plaster of paris for casting footprints, and laboratory jars for sending back hair, scat, or tissue samples to be identified by DNA analysis. We've also got video cameras to record what the migoi means to the Bhutanese themselves. Our guide is Dasho Palden Dorji, tall and chiseled, who was educated at the University of California, Santa Barbara, and speaks English like an American. He's a true believer in the migoi, but patient with skeptics. Despite his easygoing manner, locals tend to trot when he issues an order. *Dasho* is a term of respect similar to "lord," and he is a first cousin to the king.

Other Bhutanese aren't so patient. When I note the failure of numerous previous attempts to find any hard evidence for the existence of the migoi, a forestry official wryly observes that Westerners only believe a species exists if a white man has given it a Latin name. In the 1950s a "new" primate was "discovered" in southern Bhutan and named with the help of a British naturalist, E. P. Gee. The golden langur, a gregarious monkey living in groups of 15 or 16 individuals, had been well known to the Bhutanese for centuries. But it is now immortalized in scientific literature as *Presbytis geei*, a name that still rankles some people in

Bhutan. Local knowledge also got short shrift when yak herders reported seeing tigers in the mountains; science knew that tigers never go much above 6,000 feet. Then Bhutanese wildlife officials photographed a tiger crossing through a meadow at nearly 10,000 feet. So when locals who travel in the mountains say that a shy, solitary primate survives undetected there, isn't it a little arrogant to dismiss it as myth?

Maybe so, but our first few rounds of inquiry go to the skeptics. In a traditional farmhouse a few miles from the airport, a ritual purification ceremony is underway. The air is heavy with the smell of burning juniper and the sound of monks praying to drive unwanted spirits into intricate traps made of bamboo and ribbon. Dorji Tshering, the head of the household, sits cross-legged on the floor, fanning flies from his bowl of butter tea as he recalls his encounter with the migoi. The memory still gives him nightmares, though it happened 50 years ago. He and a friend had climbed the mountain behind the house in search of a suitable tree to saw into planks. Their journey took them through a glen called Migoi Shitexa and up to a shelter called Bandits' Cave. That night, as they collected firewood, they noticed strange footprints in the fresh snow. Tshering, now in his 80s, bowlegged, with a shock of white hair, taps his forearm at the elbow and then his knuckles, to show the length of the footprint. He knew what bear footprints looked like, he says, and this wasn't a bear. In the darkness, the two men heard the creaking and breaking of bamboo, followed by an eerie monotone call. Tshering imitates the sound. "This was definitely the migoi," he tells us. "We thought it was going to eat us up." "But what did you actually see?" I ask, and the answer is little more than a shape moving beyond the light of their fire, the size of a man, hairy, on two legs, its features indistinct in the darkness.

The stories told about the migoi are often like that, full of conviction but woefully short on detail. Or chock-full of plausible details, until a trapdoor suddenly drops open in the argument: a nomad digresses knowledgeably about the differences between migoi scat, which he says mainly contains bamboo, and bear scat, which is full of acorns. Then he adds that if you happen to go into the mountains when you are spiritually unclean, the migoi will bring typhoons and hailstorms down on your head. This odd mix of being so savvy about the natural world and yet so credulous is a little hard for outsiders like us to fathom. When we ask Tshering what "Migoi Shitexa" means, he says, "the place where the yeti scratches for lice." It is normal in Bhutan to be earthy and other-worldly at the same time.

The purification ceremony ends late in the day when the children carry the spirit traps out of the house and into the fields. The traps are laden with bread, fruit, money, and strands of fabric to placate the evil spirits now bottled up within. "Ghost busters for Buddhists," someone remarks. But in truth, the whole scene feels as if we've been set down in the middle of a Brueghel painting, in the Europe of 500 years ago: blue smoke in the air, a neat golden stack of rice straw in the farmyard with a straw finial on top, a pervasive sense of religious faith, and a kind of ribald peasant contentment with the course of life. Standing outside, we can hear the monks raucously chanting their last few prayers. One of them turns his weathered face over his shoulder, grinning and shouting the words of the prayer out to us through a tiny arched window. A few minutes later, a mangy dog comes trotting back to the house from the field, a strand of ribbon trailing over its ears from its raid on the contents of a spirit trap.

Face-to-Face

A winding, one-lane mountain road takes us into central Bhutan, where the monastery called Gangtey Gompa is famous for two things: black-necked cranes foraging in the broad wetlands below, and within the walls of the monastery itself a mummified *mechume*, the putative remains of a small yeti.

The black-necked cranes are real enough, and by late October their *trum-trum* calls echo across the frosted marshes. In the brilliant white light of dawn, groups of them soar down, the sun glinting off their 6-foot wingspans. In a country where the human life expectancy is about 50 years, the cranes live to be nearly half that and are revered as bodhisattva, or Buddhist deities. It's said that at the time of their departure each year, before their migration north to the Tibetan plateau, the cranes fly three times around the monastery, the clockwise ritual of any Buddhist pilgrim. There's a simpler explanation: ornithologists say the cranes are simply flying around trying to gain altitude. But for us, the symbolic and supernatural values are starting to become more intriguing. Coming from a world where we see the landscape in terms of lots and subdivisions, we're learning to envy the way people in Bhutan still tell stories about their own hills and valleys.

It's easy to get caught up in local values, especially within the darkened monastery itself, where the monks chant *Om ma ni pad me hung* ("Praise to the jewel in the lotus"), as they have chanted numberless times since the twelfth century. Our guide, Dasho Palden, says no Westerner has ever entered the monastery's inner sanctum before, but the abbot has arranged this visit especially to show us the mechume. The atmosphere in the prayer room is somewhere between a reliquary and a rag-and-bone shop. By the flickering butter candles, we step gingerly around sacks hanging like punching bags from the rafters. Asked what's in them, Palden replies, "Diseases. If they get out, they will spread." The wide, wooden floorboards have been polished by generations of barefoot pilgrims prostrating themselves before the crowded altar. Dasho Palden prostrates himself, too, and then, according to custom, a monk hands him three dice to foresee the fate of the expedition we are about to undertake into the perilous high country. Palden rolls a 13. In the Eastern scheme of things, he assures us, this is a good, solid number.

Afterward, he leads us over to the far side of the altar, to examine the mechume. The story goes that roughly 200 years ago, in a village two days' hike from here, a series of killings occurred. A local holy man determined that this mechume was the culprit. He tracked it down and cut it in half with his sword, and the corpse has been hanging at Gangtey Gompa ever since. The head, no larger than a child's, slumps down, chin resting on the sack of skin that was once its chest. The withered hands and feet hang by threads and bits of dried flesh. The eyes are squeezed shut, and the mouth is stretched wide. Dasho Palden suggests that the mouth is making the sound of the mechume, *Woooooo*. But to me it looks as if the mouth has been frozen at the split second of this creature's death in an eternal wail of agony and despair. Either way, someone has stuffed an offering of *ngultrum*, the local currency, into its mouth.

That face continues to haunt us for days afterward, and one night around the campfire, we skeptics in the group develop an alternative theory about how it came to be at Gangtey Gompa: Imagine a rural holy man beset by angry, terrified villagers demanding action. Was it practical to think he could have caught and killed a creature as elusive as the mechume, which the Bhutanese themselves sometimes describe as part god, part devil? Or did he simply find some poor scapegoat, a loner, a madman, someone who would not be missed—much as New Englanders once burned ordinary women as witches? In any case, we agree that what is hanging at Gangtey Gompa is unmistakably, unbearably, human.

Go Down the Cliff

"You'll have to excuse me," Kunzang Choden says, when she greets us at the front door of her family compound, "but I am expecting my uncle's reincarnation. He's an American boy, 14 now." He's due to arrive sometime before nightfall, and everyone's bustling around getting ready. Ugyen Choling Palace, where Choden's ancestors have lived since the fifteenth century, is at the end of the long, idyllic Tang Valley in central Bhutan. It's two hours by car from the nearest town, Bumthang, and a one-hour hike uphill from the nearest road. The palace once lorded over the valley, but almost all the rooms are empty now, with the wooden window panels drawn shut to keep out pigeons. Choden, now middle-aged, grew up before electricity, television, and videotapes came to Bhutan, and the greatest pleasure of her childhood was listening to yak herders, back from the wild, telling their stories of the high country. She's now Bhutan's leading folklorist, and she's put together an anthology called *Bhutanese Tales of the Yeti*.

Choden accepts my impression that the migoi stories are much like European fairy tales. The migoi, like an ogre in a fairy tale, often comes to a gruesome end. "I also have my sympathies with the migoi," Choden admits. "I'm sure if the migoi could tell its story, these terrible things wouldn't happen." But humans tell the stories, and "even though we Bhutanese live in nature, we have to be able to be the masters somehow."

Is it possible that's why Bhutan still has its migoi stories? Because there's still wilderness, unmastered, just beyond the farmyard gate, much as there was dark forest in Europe back when the fairy-tale tradition there was strong? Choden replies that urbanized Western countries still have their credulous tales, only updated a little: "*Star Wars* and all that comes from the fact that they've lost the wilderness and they have to look beyond, whereas we still have our environment intact, and we know that there are spirits living under the trees, spirits residing on the mountaintops. So this is still a part of our reality."

As we talk, her pet Lhasa apso noses around. "Ignore him," she advises. "He is a most terrible attention-seeker." And then she continues, "The supernatural and the natural, we do not delineate. People in the villages still perform rituals to appease the spirits that they have harmed, knowingly or unknowingly. We have, I guess, a ritual world. We have so much that is not explained. And we do not want it explained. What I fear most is that soon, with our children all going into Western schools and learning more about Western culture and beliefs, this will be lost."

Then, since our expedition is about to head off into the mountains, seeking explanations and evidence, she offers some parting advice: if we meet a female yeti, keep in mind that the yeti's long, sagging breasts make her top-heavy on a down slope. "So, yes, run downhill," Choden says. Then she grins and adds, "Go down the cliff, I think."

Next night, when our packhorses have been set free to graze, and our tents are staked down, the group gathers for dinner. We're in bear and tiger country now, our evolutionary biologist reminds us, and he gives each of us an "attack alarm." "It will produce a 107-decibel shriek, and that should scare off just about any animal." Then he pushes the trigger.

"The horses didn't twitch," someone says, when we have recovered enough to hear them contentedly tearing up the grass just beyond the light of our lanterns. We decide to put our faith instead in one of our guides, known simply as "303," for the battered old .303 Enfield rifle he keeps slung over one shoulder. The government has provided us with this armed protection because unpredictable Himalayan black bears often maul or kill unlucky yak herders.

We camp in an open meadow, but there's still forest all around. On the big old spruce trees, Spanish moss drifts in the breeze like an old man's wispy beard. Rhododendrons with leaves as big as fans grow in the understory, along with stunted little bamboo. Soft green moss carpets the floor and swells up in pom-pom clumps over broken branches. The forest itself feels as though animals—and spirits, too—have somehow taken root in the steep mountain soil. But our camera traps, with motion sensors and infrared gear for shooting by night as well as by day, turn up no big, dangerous animals. The movement of the vegetation sets them off. And once, some creature leaves its tantalizing shadow across a corner of the video frame.

Proving a species exists is relatively straightforward. The rule of habeas corpus applies: you must present a body, a type specimen that other scientists can examine in a laboratory setting. Proving that a species doesn't exist, on the other hand, is near to impossible. This dawns on us, literally, at Rodong La, at 14,000 feet on the trek from Bumthang to Lhuentse.

According to Kunzang Choden, this lonely pass is one of the most frequent sites for yeti spottings. So one night the technical wiz and I, the skeptics in our group, camp out alone at Rodong La, without 303 or any other protection, offering ourselves to the yeti as unbelievers. We spend the first part of the night holed up in a blind we've pitched in a shadowy stand of gnarled rhododendrons. It's cold and lonely. A cloud envelops us, and we have no moon or stars to give us the minimal light needed for the image intensifier on our video camera. Everything is damp and eerily silent, with no birds or insects. It feels as if the cloud has somehow sucked the sound out of the sky. On this remote spot, one of Choden's informants once saw a yeti face-to-face with a tiger.

We have no such luck. But when the lichen and moss have gone brittle with frost, the cloud lifts. The faint yellow light before dawn unveils the landscape around us, and then other hills and mountains, which gradually separate from one another as the sun rises. In the distance, brilliant orange light catches on the snow-capped peak of Gangkar Punsum. At almost 25,000 feet, it is the highest mountain in Bhutan, and no one has ever climbed it.

By now, we are out of the blind wandering in amazement. To the northeast, a wide lazy river of clouds flows through a valley and, at the end, the clouds spill down over a precipice like a waterfall, tumbling and steaming between two spruce promontories. Beyond that, other valleys, still forested and without people, delve toward infinity. It makes me think that in this vast unexplored landscape anything is possible.

It's Not About the DNA

But not necessarily right now. In all our climbing up lonely mountain passes and down endless winding stairways built into the cliff face, we turn up nothing remotely like a yeti. The camera traps yield one squirrel, and a yak herder urinating by the side of the trail. Then one day we make the steep, sweaty descent into a valley, where the white noise of a river rises up to meet us, along with the cries of farm children. In the remote, roadless village of Khaine Lhakhang, we meet a man named Sonam Dhendup, with black hair just starting to go gray, a wispy goatee, and short, rough, muscular legs—the product of a lifetime in the Himalayas.

Dhendup tells us he has worked for the past 12 years as a migoi-spotter for the government. He has yet to see a migoi himself. But he knows a place two days' hike away, where the migoi comes to eat bamboo each spring and where his droppings pile up in heaps. The season is wrong. But the mountains are tempting, and Dhendup tells us enough to convince us he knows the local wildlife.

Accompanied by 3o3, two of our party follow Dhendup back up until they are walking among the clouds at 12,ooo feet. In the dense, sodden forest there, Dhendup kneels to indicate the fresh pugmarks of a large male tiger, a few minutes ahead. The worn bolt on the .3o3 Enfield slides home with a sharp click. A soft rain begins to fall. As Dhendup creeps barefoot through the forest, he points out more tiger prints, some wild boar grubbings, places where bears have clearly rubbed against trees. He's a connoisseur of Himalayan wildlife, able to distinguish black bear and brown bear by their paw marks alone. Why would he, as skeptics like to suggest, mistake either species for a yeti?

The hill fog closes in around them, and the bamboo forest becomes a dripping prison of crisscrossed stems and pungent leaf mold. They circle aimlessly for hours, until Dhendup veers off purposefully and they arrive at the hollow tree he has been seeking. As expected, the dung heaps he saw there last spring are long gone. But there's a cavity in the tree just large enough for a human, or some creature of similar size, to hunker down and find refuge. A careful inspection turns up hairs stuck in the rough sides of the cavity. Some of the strands are long and dark, others shorter and more bristly. Not bear, says Dhendup. The hairs are, at the very least, worth packing up and taking home for DNA analysis.

Months later, in a laboratory at Oxford University, geneticist Bryan Sykes is pleased with the samples we have brought back. The hairs have plump follicles, the part that contains the DNA. Prospects for identifying the source of these hairs seem good. Sykes starts with the plausible assumption that he is looking at bear DNA, or maybe wild pig. But the DNA seems to suggest otherwise. His laboratory simply cannot sequence it. "We normally wouldn't have any difficulty at all," says Sykes. "It had all the hallmarks of good material. It's not a human, it's not a bear, nor anything else that we've so far been able to identify. We've never encountered any DNA that we couldn't sequence before. But then, we weren't looking for the yeti." It is, he says, "a mystery, and I didn't think this would end in a mystery."

What a scientist cannot add is that sometimes a mystery is enough. I at least had conceded as much that morning back in the mountain pass called Rodong La. Watching the sunrise on the endless mountains, it seemed to me that if I had a choice—a land rich in wilderness, and rich in demons, too, or a land "civilized" and full of skeptics like me—I would gladly take Bhutan the way it is, yeti and all.

Afterword

I was back in Africa on an assignment when my forty-eighth birthday arrived, a big one for me, as I had been living for the previous 25 years with a stupid premonition that I would die at the age of 47. So I had been feeling a little edgy at times over the past year as I swam in piranha-infested rivers and traveled in the company of lions, leopards, Cape buffalo, and television camera crews (not to mention the odd assignment driving in a demolition derby). It might sound strange that I would head off to Africa while expecting imminently to die, but I figured I could die just as easily on a highway back home.

Then of course it didn't happen, and I felt like one of those idiots with a poster saying, "The end is nigh," when it isn't. Back at the lodge on my last night as a 47-year-old, I celebrated with several bottles of lager. A glass of Talisker seemed like a good idea for a nightcap, in honor of certain animals and the scientists who study them. Then I said goodnight and staggered out onto the trail back to my tent. Alone in the dark, it occurred to me that, given the difference in time zones, I wouldn't actually turn 48 till around 10:30 the next morning. And then I heard the sound of an animal nearby.

People have this nutty idea that hippos are those cute things in tutus dancing in Disney films or doing the macarena on You-Tube. But think of yourself as the dance floor. If you get between them and the water in the dark, when they're busily tearing up grass to eat, hippos tend to panic and then they trample or gore you. A colorful way to go, all guts, no glory. But not that night. I made it back to my tent intact and slept peacefully.

Next morning, we drove through a stand of acacia bush and found ourselves in the middle of a large group of elephants. They shook their heads at us and shrieked, no more than 20 or 30 feet from where I sat on top of the Land Rover. But it was all talk. A birthday greeting. I was ecstatic to be alive, and to be in such a world, among such creatures. I threw my head back and trumpeted with them.

Then we turned and headed on into the bush.

Acknowledgments

I am deeply thankful to the John Simon Guggenheim Memorial Foundation for their support and encouragement. I am also grateful to the many readers who have sent their comments and criticisms (even the ones who tear out pages and cover them with handwritten annotations on my stupidity). I'd like to thank Harry Marshall of Icon Films who contributed considerable thought and planning, as well as some eloquent wording, to the chapter on Bhutan. Also at Icon, thanks to Laura Marshall, who kept me warm at night in the Himalayas (that sleeping bag liner really worked!). At *Smithsonian* magazine, I'd like to thank Jim Doherty, a writer's editor, who got me raises without my having to ask and sent frequent notes of insult and praise, pounded out on yellow paper with a battered old Royal standard typewriter. Doherty and Bob Poole at *National Geographic* also had the uncommon sense not to dither endlessly before making an assignment. At times, a word was sufficient: "Conniff. Leopards. Go." At *Smithsonian*, thanks also to Don Moser, Carey Winfrey, and Laura Helmuth for their editing; Beth Py-Lieberman and Helen Starkweather for their help getting the facts right; and Andrea Georgiou for many favors. At *National Geographic* magazine, I am grateful to editors Jennifer Reek, Chris Johns, Oliver Payne, and Carol B. Lutyk, and also to meticulous researchers

Eileen Yam, Jennifer Fox, Carolyn H. Anderson, and Kathy B. Maher. My thanks to Steve Petranek, David Grogan and Jessica Ruvinsky at *Discover* magazine, Mary Turner at *Outside* magazine, David Seideman and Rene Ebersole at *Audubon* magazine, Mel Allen at *Yankee* magazine, David Hamman, Helene Heldring, guide Newman Chuma and others at Chitabe Camp in Botswana. For their considerable help, my thanks to agent John Thornton, and to my editor Angela von der Lippe at W. W. Norton. Finally, I give thanks and much love to my family for acerbic comments and other forms of help, particularly to my father James C. G. Conniff, my siblings Greg Conniff and Cynthia Cavnar, my wife Karen and our children James, Ben, and Clare.

Permissions

Smart Reading for Doing Dumb Things

Wild Dogs   **p. 16**

Boggs, L., and J. McNutt. 1997. *Running wild: Dispelling the myths of the African wild dog*. Washington, DC: Smithsonian Books.

Creel, N., and S. Creel. 2002. *The African wild dog: Behavior, ecology, and conservation (Monographs in behavior and ecology)*. Princeton: Princeton University Press.

Life on the Web   **p. 33**

Eberhard, W. G. 2000. "Spider manipulation by a wasp larva," *Nature* 406 (6793): 255–56.

Foelix, R. 1996. *Biology of spiders*. 2nd ed. New York: Oxford University Press.

Wise, D. 1995. *Spiders in ecological webs (Cambridge studies in ecology)*. New York: Cambridge University Press.

The Value of a Good Name   p. 44

Epstein, M. E. and P. M. Henson. 1992. "Digging for Dyar: The man behind the myth." *American Entomologist* 38:148–69.

Erwin, T. 1988. "The tropical forest canopy: The heart of biotic diversity." In *Biodiversity*, ed. by E. O.Wilson. Washington, DC: National Academy Press.

——. "Biodiversity at its utmost: Tropical forest beetles." In *Biodiversity II*, ed. by M. L. Reaka-Kudla, D. E. Wilson, and E. O.Wilson. Washington, DC: Joseph Henry Press.

Isaak, M. "Curiosities of biological nomenclature," at http://home .earthlink.net/~misaak/taxonomy.html.

Menke, A. "Fun New York or curious zoological names," at http://cache .ucr.edu/~heraty/menke.html.

Yanega, D. "Curious scientific names," at http://cache.ucr.edu/~heraty/ yanega.html.

Lemurs in Love   p. 52

Jolly, A. 2004. *Lords and lemurs: Mad scientists, kings with spears, and the survival of diversity in Madagascar*. Boston: Houghton Mifflin.

Wright, P. C. 1995. "Demography and life history of free-ranging *Propithecus diadema edwardsi* in Ranomafana National Park, Madagascar." *International Journal of Primatology* 16(5):835–54.

——. 1999. "Lemur traits and Madagascar ecology: Coping with an island environment." *Yearbook of Physical Anthropology* 42:31–72.

Bluebloods   p. 70

Shuster, C., J. Brockman, R. B. Barlow, eds. 2004. *The American horseshoe crab*. Cambridge: Harvard University Press.

The King of Pain   **p. 80**

Evans, D. L. and J. O. Schmidt. 1990. *Insect defenses: Adaptive mechanisms and strategies of prey and predators*. New York: State University of New York Press.

Schmidt, J. O. 1988. *Insect venoms*. Westport, CT: Praeger Publishers.

Life List   **p. 94**

Farris-Toussaint, L., and B. De Wetter. 2000. *On the trail of monkeys and apes*. New York: Barrons Educational Series.

Mittermeier, R. 2006. *Lemurs of Madagascar* (Tropical field guides). Washington, DC: Conservation International.

Ghosts in the Grasslands   **p. 101**

Durant, S. 1998. "Competition refuges and coexistence: An example from Serengeti carnivores." *Journal of Animal Ecology* 67(3):370–86.

Durant, S. 2000. "Living with the enemy: Avoidance of hyenas and lions by cheetahs in the Serengeti." *Behavioral Ecology* 11(6):624–32.

Hunter, L., and D. Hamman. 2007. *Cheetah*. Johannesburg: Struik.

Kelly M., M. Laurenson, et al. 1998. "Demography of the Serengeti cheetah (*Acinonyx jubatus*) population: The first 25 years." *Journal of Zoology* 244: 473–88.

O'Brien, S., M. Roelke, et al. 1985. "Genetic basis for species vulnerability in the cheetah." *Science* 227(4693):1428–34.

The Enemy Within   **p. 121**

Abe T., D. Bignell, M. Higashi, eds. 2000. *Termites: Evolution, sociality, symbioses, ecology*. New York: Springer.

Marais, E. N. 1973. *The soul of the white ant*. London: Penguin Books.

Pimentel, D., et al. 2000. "Environmental and economic costs of non-indigenous species in the United States." *BioScience* 50(1):53–65.

The Monkey Mind   **p. 135**

Cheney, D. L., and R. M. Seyfarth. 1992. *How monkeys see the world: Inside the mind of another species*. Chicago: University of Chicago Press.

———. 2008. *Baboon metaphysics: The evolution of a social mind*. Chicago: University of Chicago Press.

Strum, S. C. 2001. *Almost human: A Journey into the world of baboons*. Chicago: University of Chicago Press.

Hummers   **p. 148**

Stokes, L., and D. Stokes. 1989. *Stokes hummingbird book: The complete guide to attracting, identifying, and enjoying hummingbirds*. London: Little, Brown and Company.

Williamson, S. 2002. *A field guide to hummingbirds of North America* (Peterson field guide series). Boston: Houghton Mifflin.

Family Politics   **p. 165**

Dawkins, R. 1979. *The selfish gene*. London: Oxford University Press.

De Waal, F. 1982. *Chimpanzee politics*. New York: Harper Publishing.

———. *Peacemaking among primates*. Cambridge: Harvard University Press.

———. *Good natured: The origins of right and wrong in humans and other animals*. Cambridge: Harvard University Press.

———. *My family album: Thirty years of primate photography*. Berkeley: University of California Press.

Lorenz, K. 2002. *On aggression* (Routledge classics). New York: Routledge.

Machiavelli, N. 2003. *The prince* (Penguin classics). London: Penguin Classics.

A Little Sneaky Sex   **p. 179**

Birkhead, T. R., and A. P. Møller, eds. 1998. *Sperm competition and sexual selection*. London: Academic Press.

Shuster, S. M. 2002. "Mating strategies, alternative." In *Encyclopedia of Evolution*, ed. by M. Pagel, et al. Oxford: Oxford University Press, 688–93.

Sinervo B., C. M. Lively. 1996. "The rock-paper-scissors game and the evolution of alternative male strategies." *Nature* 380(6571):240–43.

Swamp Thing   **p. 190**

Pritchard, P.C.H. 2006. *The alligator snapping turtle: Biology and conservation*. Malabar, FL: Krieger Publishing Company.

Whitfield G., A. C. Steyermark, M. S. Finkler, and R. J. Brooks, eds. 2008. *Biology of the snapping turtle* (Chelydra serpentina). Baltimore: The Johns Hopkins University Press.

Backyard Wildlife   **p. 200**

Conant, R. et al. 1998. *A field guide to reptiles & amphibians of Eastern & Central North America* (Peterson field guide series). Boston: Houghton Mifflin.

Covell, C. V., and R. T. Peterson. 1984. *Peterson field guide to eastern moths* (Peterson field guide series). Boston: Houghton Mifflin.

Hurd, P. D. 1954. "Myiasis resulting from the use of the aspirator method in the collection of insects," *Science* 119(3101):814–15.

Klots, A. B. 1951. *A field guide to the butterflies* (Peterson field guide series). Boston: Houghton Mifflin.

Mikula, R. 2001. *Garden butterflies of North America: A gallery of garden butterflies & how to attract them.* Minocqua, WI: Willow Creek Press.

National Audubon Society. 1980. *National Audubon Society field guide to North American insects and spiders* (Audubon Society field guide). New York: Knopf.

———. 1996. *National Audubon Society field guide to North American Mammals* (Revised and Expanded) (Audubon Society field guide). New York: Knopf.

National Geographic Society. 1989. *National Geographic Society field guide to the birds of North America.* Washington DC: National Geographic Society.

Reid, F. 2006. *Peterson field guide to mammals of North America. 4th. ed.* (Peterson field guide series). Boston: Houghton Mifflin.

Sibley, D. A. 2000. *The Sibley guide to birds.* New York: Knopf.

On the Track of the Cat p. 205

Hollis-Brown, L. A. 2005. "Individual variation in the antipredator behavior of captive rhesus monkeys (*Macaca mulatta*)." PhD thesis, University of California, Davis.

Stander, P. 1998. "Spoor counts as indices of large carnivore populations: The relationship between spoor frequency, sampling effort and true density" *The Journal of Applied Ecology* 35(3):378–85.

Every Ant on Earth at Your Fingertips p. 217

Cover, S. P., and B. L. Fisher. 2007. *Ants of North America: A guide to the genera*. Berkeley: University of California Press.

Fisher, B. "About AntWeb," at www.antweb.org.

Goodman, S., J. P. Benstead, and H. Schutz. 2007. *The natural history of Madagascar*. Chicago: University of Chicago Press.

Hölldobler, B., and E. O. Wilson. 1990. *The ants*. Cambridge: Harvard University Press.

Tyson, P. 2001. *The eighth continent: Life, death, and discovery in the lost world of Madagascar*. New York: Harper Perennial.

Jelly Bellies p. 232

Ivanov, V., et al. 2000. "Invasion of the Caspian Sea by the comb jellyfish *Mnemiopsis leidyi* (Ctenophora)." *Biological Invasions* 2(3):255–58.

Wrobel, D., and C. Mills. 1998. *Pacific Coast pelagic invertebrates: A guide to the common gelatinous animals*. Monterey: Monterey Bay Aquarium.

All Piranhas Want Is a Nice Piece of Tail p. 245

Haddad, V., and I. Sazima. 2003. "Piranha attacks on humans in southeast Brazil: Epidemiology, natural history, and clinical treatment, with description of a bite outbreak." *Wilderness & Environmental Medicine* 14(4):249–54.

Myers, G. 1972. *Piranha book*. Neptune City, New Jersey: TFH Publications.

Roosevelt, T. 1914. *Through the Brazilian wilderness, with illustrations by Kermit Roosevelt and other members of the expedition*. New York: Scribner's.

Sazima, I., and S. D. Guimaraes. 1987. "Scavenging on human corpses as a source for stories about man-eating piranhas." *Environmental Biology of Fishes* 20(1):75–77.

Schleser, D., 2008. *Piranhas (Complete pet owner's manual)*. New York: Barrons Educational Series.

Winemiller, K. O., and L. C. Kelso-Winemiller. 1993. "Predatory response of piranhas to alternative prey." *National Geographic Research* 9:344–57.

Intimate Friends   p. 264

Busvine, J. R. 1976. *Insects, hygiene and history*. London: Athlone Press.

Lehane, M. J. 2005. *The biology of blood-sucking in insects*. Cambridge, England: Cambridge University Press.

Rosebury, T. 1969. *Life on man*. New York: Viking Press.

Zinsser, H. 1934, 1996. *Rats, lice, and history*. New York: Black Dog & Leventhal Publishers.

In the Realm of Virtual Reality   p. 274

Bateman, R., and P. Matthiessen. 2001. *The birds of heaven: Travels with cranes*. New York: North Point Press.

Choden, K. 1994. *Folktales of Bhutan*. Bangkok: White Lotus.

——. *Bhutanese tales of the yeti*. Bangkok: White Lotus.